THIRD WORLD REGIONAL DEVELOPMENT
A Reappraisal

Edited by
DAVID SIMON

P·C·P
Paul Chapman
Publishing Ltd

First published 1990
Paul Chapman Publishing Ltd
144 Liverpool Road
London N1 1LA

British Library Cataloguing in Publication Data
Third World regional development: a reappraisal.
  1. Developing countries. Economic development
  I. Simon, David
  330.91724

ISBN 1 85396 121 3

Typeset by Inforum Typesetting, Portsmouth
Printed and bound by Athenaeum Press, Newcastle upon Tyne

A B C D E F G   6 5 4 3 2 1 0

SB 1882 £15.95. 11-90

THIRD WORLD
REGIONAL DEVELOPMENT

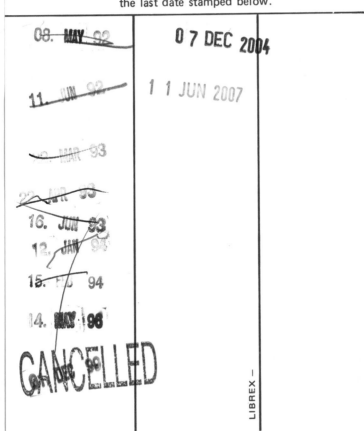

David Simon lectures in development studies and geography at Royal Holloway and Bedford New College, University of London. He is co-author of *The British Transport Industry and the European Community* (Gower, Aldershot, 1987), is currently writing a monograph on African cities in relation to the world economy and has published widely on transport, urbanization, planning and environmental aspects of development. His Third World research interests focus primarily on southern Africa and, as a Namibian specialist, he is involved in current planning initiatives for that country's development after independence.

CONTENTS

LIST OF CONTRIBUTORS

Bill Gould is senior lecturer in geography in the University of Liverpool. He has written and researched extensively on migration in Africa and on the problems of educational development at the local level in Africa and Asia. He has acted as consultant on these themes to the World Bank, UNESCO and the Aga Khan Foundation. His work in Ghana has been associated with a United Nations Development Programme/International Institute for Educational Planning/Ministry of Education and Culture project for the Training of Educational Planners for district-level responsibilities in its new decentralized structure. He is the joint editor of *Planning for Population Change* (with R. Lawton) and editor of a special issue of *Geoform* on skilled international labour migration (1989). He has written several articles based on his own recent research on the migration of schoolleavers in Kenya.

Ruud H. F. Jansen is a research co-ordinator working at the University of Utrecht, Department of Geography of Developing Countries, responsible for a rural development research project in Botswana. He worked in Botswana from 1983 to 1987 as a district land use planner in the country's Southern District.

Stella Lowder's interest in development, particularly from the Latin American standpoint, dates from experience in the family business and a spell, after graduation, in the social-survey research unit of the Ministry of Work in Lima. A geographer, she has since then conducted research on migration, colonization and urbanization and written *Inside Third World Cities* (Croom Helm, Beckenham, 1986). This chapter stems from several visits to Ecuador, the longest being nine months in 1985–6, while conducting research into urban-planning mechanisms in intermediate cities. She lectures in the Department of Geography and Topographic Science, University of Glasgow.

Roswitha Newels, currently working for the United Nations as Regional Planning Expert in the Republic of the Marshall Islands, has worked for several years with the UNCRD and the UN Department of Technical Co-operation for Development in Pacific countries. She holds a doctorate in regional and urban planning from the University of Paris IV-Sorbonne.

Carole Rakodi worked as a planner in Zambia from 1971 to 1978. Since then she has been a lecturer in the Department of Town Planning, University of Wales, College of Cardiff, and has published widely on urban planning and housing in developing countries, especially with reference to Zambia. Since 1986 she has visited Zimbabwe several times and in 1988 visited Tanzania, Zambia and Zimbabwe, with the assistance of a grant from the Economic and Social Research Committee of the Overseas Development Administration. She also attended the inaugural workshop of the Rural and Urban Planning Network of Southern and Eastern Africa in Harare in September 1988.

Arie Romein is employed by WOTRO, The Netherlands Foundation for the Advancement of Tropical Research. He is carrying out research on the absorption capacity of small urban centres, for rural migrants in the Northern Region of Costa Rica and has published two articles on rural colonization in Costa Rica.

Marcel M. E. M. Rutten is a human geographer from the Study Group for Developing Regions at the Faculty of Policy Sciences, University of Nijmegen. He has worked in the Caribbean and in Kenya and is currently conducting research in Kajiado District (Kenya) among the Maasai nomadic pastoralists. His major interests lie in the fields of rural economy, agrarian change, ecology, nomadic people, remote sensing and geo-information systems, on which he has published several articles. He was co-editor (with A. Dietz) of *The Future of Maasai Pastoralists in Kajiado District, Kenya; Integrated Proceedings of a Conference in Brackenhurst Baptist International Conference Centre* (Limuru, 1989).

Jur Schuurman works at the Foreign Relations Centre of the University of Nijmegen, The Netherlands. Before that, he was employed by the Universidad Nacional in Heredia (Costa Rica), preparing a project on regional development planning. He has published several articles on Costa Rica.

James D. Sidaway is a doctoral student in the Department of Geography at Royal Holloway and Bedford New College, University of London, working on a thesis on the territorial organization of the post-independent Mozambican state. Prior to this he was at the Graduate School of European and International Studies, Reading University, where he did an MPhil in international studies.

Hidehiko Sazanami is Director of the UN Centre for Regional Development (UNCRD), which is based in Nagoya, Japan. The centre's activities all aim at strengthening regional development, planning and management in developing countries. These encompass research, training, information dissemination and technical advisory services.

Leslie Sklair received his PhD in sociology at the London School of Economics, where he is presently Senior Lecturer in Sociology. He was a visiting research fellow at the Center for US–Mexican Studies, University of California, San Diego, in 1986–7 from where he carried out the research on the maquilas. In connection with his current research on the sociology and politics of the transnational corporations, he has acted as a consultant to the UN Centre on Transnational Corporations. His recent publications on this work include *Assembling for Development: The Maquila Industry in Mexico and the United States* (Unwin Hyman, Boston, Mass., 1989).

Peter M. Slowe is a senior lecturer in geography at the West Sussex Institute of Higher Education. He obtained his doctorate in regional development at Nuffield College, Oxford University, in 1978. Between 1978 and 1984 he worked in the transport industry in France and Switzerland. Since then he has developed an interest in the relationship between political philosophy and geographical factors, as reflected in recent work on the Sino–Soviet boundary, Quebecois separatism, Berlin 1945 and West Africa. His most recent publication is *Geography and Political Power* (Routledge, London, 1990).

Jan J. Sterkenburg is an assistant professor in the Department of Geography of Developing Countries, Faculty of Geographical Sciences, University of Utrecht. He specializes in rural development and regional planning in Third World countries and has working experience in a number of African and Asian countries, including Kenya, Tanzania, Ghana, Botswana and Sri Lanka. Recent publications include *Agricultural Commercialization and Government Policy in Africa* (with J. Hinderink, 1987) and the paper, 'Rural development policies; cases from Africa and Asia' (Utrecht, 1987). At present he is involved in an evaluation study of The Netherlands government rural development aid programme.

Allert van den Ham studied the geography of developing areas at the Catholic University of Nijmegen, The Netherlands. He has lived in Indonesia since 1984, working as a consultant with the Regional Development Planning Boards of Aceh and West Java. Presently, he is engaged in an Indonesian–Dutch rural development project among coffee growers in the district of Central Aceh.

Paul J. M. van Hoof is staff member of the Department of Geography of Developing Countries at the University of Utrecht, The Netherlands. He studied human geography at the University of Nijmegen and specialized in rural development and environmental problems in the Third World. He gained work experience in Kenya, the People's Republic of China and Botswana. He is now working on a PhD thesis titled 'Decentralization of planning and policy for rural development in Botswana'.

Ton van Naerssen is senior lecturer in the geography of Developing Areas at the Faculty of Policy Sciences, Catholic University of Nijmegen, The Netherlands. His areas of interest are development policy and urban–regional planning in southeast Asia. He wrote his thesis on industrialization and regional development in Malaysia (1983) and was co-editor of *Urban Social Movements in the Third World* (Routledge, London, 1989).

PREFACE AND ACKNOWLEDGEMENTS

Regional development and planning have not yet fully come of age. Although the role of space in development has been increasingly acknowledged over the last twenty to thirty years, attempts at explicit spatial planning have seldom achieved significant results. Nowhere are the contrasts between wealth and poverty, and their consequences, more clearly evident than in the Third World, where inequalities at almost every scale persist or have even widened. The underlying causes and potential solutions continue to generate much debate and a voluminous literature. As argued here, the basic problems with regional planning have been an inappropriate conception of space as somehow separable from other dimensions of society, economy and polity, together with the lack of political will to restructure and redistribute power. We hope that this volume will contribute to a greater understanding of recent conceptual advances and real-world experiences in this regard, and consequently to the implementation of policies more appropriate to the empowerment of the poor.

This book has its origins in contacts between British and Dutch geographers undertaking research and consultancy on regional issues in various Third World countries. We met for the first time in Amsterdam in April 1985, under the auspices of the Developing Areas Research Group of the Institute of British Geographers (IBG) and the Developing Countries Section of the Koninklijk Nederlands Aardrijkskundig Genootschap (KNAG - Royal Dutch Geographical Society). A selection of the papers presented there, on the theme of regional inequality, migration and development, appeared in a special issue of the journal, *Tijdschrift voor Economische en Sociale Geografie* (Vol. 77, no. 1, 1986).

For the second symposium, held on 6–7 April 1989 and out of which this book arises, the focus shifted from regional problems to potential solutions through an examination of existing regional development strategies and attempts to mitigate the impact of the debt crisis and structural adjustment. The symposium was hosted by the Centre for Developing Areas Research (CEDAR) within the Department of Geography, Royal Holloway and Bedford New College, University of London. Once again, we are indebted to the IBG and KNAG, without whose financial assistance the meeting would not have been possible. The meeting was also held in association with the International Geographical Union Commission on Third World Development. Although unable to attend the symposium, the United Nations Centre for Regional Development also contributed a paper on their work.

In producing this volume, the intention has been to provide coherent treatment of key themes in relation to contemporary regional development and planning, rather than simply a set of symposium proceedings. To this end, contributors willingly agreed to make sometimes significant changes to their original papers

to a tight production schedule and in line with guidelines and suggestions from
a persistent editor. Moreover, for many, English is not their first language. The
contributors' enthusiasm and the encouragement I received have been remarkable,
and are hopefully reflected in the final product. Consequently, it has been possible
to include all but a few of the papers presented April 1989, providing coverage
of an unusually broad range of country experiences to inform the conceptual
analyses. Further, some issues raised in discussion at the symposium are reflected
in the introductory and concluding chapters.

I owe debts of gratitude to the contributors, especially those who, together with
Rob Potter, commented helpfully on drafts of my work, and to Marianne Lagrange
and Catherine Blishen of Paul Chapman Publishing, for their confidence in this
venture and assistance throughout. Collectively, we are grateful to the cartographers
and secretaries in our respective institutions for their work behind the scenes.

David Simon
Egham, Surrey
March 1990

INTRODUCTION

David Simon

Regional development and spatial planning acquired widespread legitimacy during the late 1960s and 1970s, in no small measure due to the 'scientific' mantle of the relatively new discipline of regional science, which first emerged in the late 1950s. The 'historical moment' was also appropriate, since many problems had been experienced with earlier development strategies that failed to differentiate across space or to take account of the spacial impacts of economic and political forces.

Much evidence subsequently emerged, however, to suggest that incorporation of an explicitly spatial planning component did not in itself prove significantly more effective. Although greater resources were often being devoted to strategies aimed at dispersing development within national territories, such benefits as accrued all too often remained localized.

Over the last decade, in particular, social scientists have made great strides in understanding the reasons for such failure. The answer seems to lie in the way in which we treat space in relation to economic, political and social forces and factors. At a practical level, planners have endeavoured to fine-tune their policies for greater effectiveness but, although some progress has undoubtedly been made, they have not in the main yet absorbed the conceptual advances. The global political and economic system has also undergone profound changes during the 1980s, with the debt crisis and its implications looming large over all development efforts in the Third World. As the 1990s begin, the pace and significance of change are accelerating. Against this background, then, there is a widely felt need to re-evaluate the practice of regional development planning, taking stock of recent trends and seeking to bridge the gap between theory and practice. That is the objective of this book. Although the focus is specifically on the Third World, the principles and broad issues are obviously of general applicability.

Definitions

It is as well to clarify at the outset how we define the terms 'Third World', 'region', 'development' and 'planning', since these are central to the ideas developed throughout this book. The Third World is taken to include those countries outside the advanced capitalist states of western Europe, North America and Australasia (the 'First World') and the hitherto centrally planned economies of the USSR and Eastern Europe (the 'Second World'). In other words, we are focusing on that rather heterogeneous group of countries that was exploited and did not become powerful during the process by which the current world order was established. Inevitably, such shorthand labels are contentious, and there is a well developed literature on the advantages and drawbacks of these and the alternatives. Diversity within each of these 'worlds' is certainly increasing, and the recent dramatic changes in political orientations and economic systems around the globe (nowhere more conspicuously than in Eastern Europe) are likely to necessitate a review of terminology in due course. Nevertheless, the term Third World is not used here in a pejorative sense or to denote separateness and the lack of linkages with the First and Second Worlds. On the contrary, the definition cited above focuses on historical process and the fundamental but inequitable linkages between parts of the global system. Probably the best elaboration of this view is that of Abdallah (1978). It is in this sense that the term has been used as a vehicle for promoting unity and joint action in search of a new international order by leaders of the Third World. An understanding of the process by which contemporary structural problems arose is essential to the task of more appropriate development planning as set out in this volume.

By a region we understand simply a subnational division of space, delimited in terms of one or more criteria. Whatever these criteria, the delimitations reflect physical and/or sociospatial diversity. Usually, but not necessarily, regions are defined by the state so as to form the focus of lower-level administration or planning. Self-delimitation (e.g. of the area traditionally controlled by a particular group) is also possible. There is no unique way of delimiting regions, which can exist at any scale, although they generally denote one or more intermediate hierarchical levels between national and local administrations.

We take development to mean a multifaceted process whereby the quality of life and 'personality' of individuals and groups improves. This embraces social, cultural, economic and political facets, qualitative as well as quantitative variables, and is often difficult to measure empirically. It cannot be equated with purely economic development.

Planning is an activity of the state apparatus to co-ordinate, rationalize and/or (re-)organize human activity and the distribution of resources. The definition *per se* says nothing about the motives for such activities – they could be genuinely concerned with promoting development, or at least in part self-serving in terms of legitimizing existing power relations and state structures.

Organization of the book

The contents of this volume have been arranged so as to highlight the different approaches to spatial planning adopted during the last ten to fifteen years, together with key theoretical debates and salient issues that have emerged over that period. This is reflected in the division of the book into four parts, successively containing broad conceptual overviews and analytical frameworks, experiences with different selective investment strategies, a variety of decentralized approaches to development planning and conclusions and prospects. However, these are linked by several unifying themes, reflecting conceptual issues raised in Chapter 1. Foremost among them is a clear understanding that spatial dimensions cannot meaningfully be analysed in isolation from other components of society, economy and polity. For example, it is necessary to examine regional development and policy in the context of the nature of the individual state. The impacts of recession, the debt crisis and structural adjustment are also evident almost everywhere. In order to highlight these crucial and common issues, the structure just described is deemed more appropriate than dividing the book on the basis of geographical regions.

In the first chapter, David Simon surveys alternative approaches to regional development and planning, together with their theoretical underpinnings, explaining how our conceptual understanding of the issues has advanced over the last fifteen years or so. Coverage embraces both conventional theories based on neoclassical economic principles and political economy alternatives, distinguishing top down and bottom up strategies in each category, and focusing on the conceptualization of space in both theory and practice. Events during the period under review have renewed attention on devolution of decision-making and subnational planning as a response to the perceived shortcomings of centralized regimes across the ideological spectrum. Planning as a purely technical exercise has also been increasingly discredited, and the essentially political nature of planning interventions concomitantly recognized. One consequence of these developments is the formulation of more robust perspectives and planning strategies that have the potential to contribute towards improved living conditions for, and the empowerment of, the poor in the Third World.

In Chapter 2, James Sidaway and David Simon pursue some of these issues by examining uneven development and spatial policies in 'Marxist-Leninist' states. The dearth of literature on the subject is all the more remarkable, given the number of states professing adherence to some or other variant of socialism. Some of the arguments raised earlier in Chapter 1 are developed, suggesting that generalizations in some recent writings fail adequately to take account of the significant differences in ideology and policy within this category of states. The chapter then moves on from definitional questions to examine the actions of Third World socialist states in space. Of late, such regimes have commonly suffered severe pressures both from within and without, hence in many cases the sole preoccupation in terms of spatial policy has been securing territorial integrity. More generally, however, analysis of the overall organization of the state in space is likely to be more fruitful than a narrow concern with regional policy. The example of supposed anti-urban bias in early post-revolutionary societies, and postulated changes in state attitudes to urbanization and uneven development during successive stages of a transition to socialism, confirms the diversity of real-world experience. Moreover, state ideology may not be hegemonic: alternative visions of how a socialist state should be organized sometimes arise within different regions of a single country. Finally, greater account needs to be taken of non-capitalist Third World experiences in the formulation of theories about uneven development and global restructuring.

Peter Slowe's study of Guinea in Chapter 3 clearly illustrates the impact of changing official ideologies on organization of the state in space, by contrasting the Africanist national-integration approach pursued by Sékou Touré, with the open-economic, capitalist-growth strategy adopted under the military government of Lansana Conté since Touré's demise in 1984. The differences are examined under the headings of political, ethnic and class integration. In order to promote national integration in a country of distinct ethnoregional identities, and minimize class differentiation, Touré abolished chieftainships, excluded the capitalist classes and former chiefs from ranking positions in the ruling party and ensured party hegemony over all aspects of life. An effective grassroots feedback system ensured that the central state remained in touch with local opinion, and devolution of decision-making over local affairs to village level was undertaken as a precaution against the concentration of all power in central-party bureaucratic hands. Conté's regime immediately disbanded the party and village assemblies, removed economic controls and restrictions and encouraged foreign investment and greater integration into the international economy. The regime has to date relied on the strong inherited sense of national identity, rather than instituting any direct measures, to avoid the serious potential ethnoregional and political consequences of free market economics.

In Chapter 4, Hidehiko Sazanami and Roswitha Newels provide insights into very different state structures and subnational planning issues in the Pacific island microstates. Fundamental development constraints include limitations of local resources and personnel, small and often dispersed populations, extremely peripheral and dependent positions and roles within the world economy, and heavy reliance on external assistance. Uneven development at a very small scale and the need for popular participation are brought sharply into focus. Internationally co-ordinated initiatives by the United Nations and other agencies have consequently assumed great importance. In outlining the approach of the United Nations Centre for Regional Development (UNCRD), the authors provide important insights into practical analytical frameworks being instituted in situations where co-ordinated subnational development planning is in its infancy. They argue that decentralization *per se* has proved ineffectual or even counterproductive in the absence of appropriate integrative frameworks, and advocate multilevel development planning and management that incorporates an explicit spatial element and permits maximum local participation. The reader will detect some differences of perspective from that of many other contributors to the volume, who advocate reduced emphasis on specifically spatial planning in favour of more sensitive planning to exploit the spatial impress of macro-economic and sectoral policies.

Part II, on selective investment strategies, opens with Chapter 5, a study by Stella Lowder of regional policy in Ecuador. It focuses both sectorally (on agrarian and industrial strategy) and spatially (on the impact these have had on intermediate cities). During the 1970s, oil revenues provided the stimulus for a large scale modernization programme. This provides a classic example of surplus extraction from agriculture in favour of consumer-oriented industrialization centred on the cities. Agricultural exports rose but food production declined and small producers lost out to large commercial operations. State policies did bring social benefits, but strongly favoured the middle classes and élite overall. Class differentiation and urban–rural disparities have widened, especially since regional policy initiatives were focused on commercial and urban production. When oil prices fell, foreign borrowing was used to support continuation of the programmes, but by 1982 the country's external debt had reached critical proportions. Under structural adjustment, the poor in both rural and urban areas are again suffering disproportionately.

In Chapter 6, Arie Romein and Jur Schuurman examine Costa Rica's regional development strategy and its impact on the Northern Region. They argue that, although the policy has had twin objectives (namely, reducing interregional disparities and incorporating remote regions into the export-oriented commercial economy), only the latter has been seriously pursued. The primary policy instrument has been investment in large scale cattle raising, of benefit exclusively to the capitalist élite and North American consumers. Policy has always been highly centralized, and has had a negative impact on the majority of the regional population and upon the environment. Moreover, little reduction in regional disparities occurred. With the onset of the debt crisis and increasing pressure to raise export levels, the objective of reducing regional inequality was all but abandoned. Nevertheless, they conclude that there has not been a noticeable weakening of the territorial component of regional planning in favour of the functional component because the former has never really existed.

A rather different type of strategy, export-led industrialization through the deployment of foreign investment and technology, forms the subject of Chapter 7. Leslie Sklair compares the experiences of export-processing zones in Mexico (the maquilas along the US border) and China (Shenzhen Special Economic Zone). These regional policy initiatives symbolize the pressures for foreign investment, employment creation in peripheral regions and closer incorporation into the global capitalist economy by states of very different character. In each case, the impact of the zone on development is evaluated against a set of six criteria. Although economic growth has occurred, the positive developmental impacts have been very limited because of conflicts of interest between the local working classes and transnational capital.

In Part III, the focus shifts to recent strategies aimed at decentralized, predominantly rural, development. In Chapter 8, Carole Rakodi provides an interesting comparative study of subnational planning in Tanzania, Zambia and Zimbabwe. Given their predominantly rural population distributions, area-based rural development strategies receive special attention. Despite differences in state ideology and organization, all three have been characterized by centralized decision-making since independence. The governments of all three have also shared an ambivalence towards the prospect of a prosperous and politically influential peasantry. The extent of population relocation and sociospatial reorganization under Tanzania's *ujamaa* villagization scheme is distinctive, but in practice, as in the other countries, collectivization of agriculture never embraced more than a small proportion of peasants. More recently, the debt crisis and IMF prescriptions have forced the adoption of increasingly market-oriented policies. Particularly in Zambia, the policies of other donors have often led to a proliferation of divergent agricultural policies within and between regions, with obvious implications for the coherence of subregional planning. Rakodi also examines national urban development strategies, the central plank of regional development planning in all three countries, which have shared a concern that major cities have been growing excessively rapidly. An essentially physical approach to planning, without due understanding of the associated political and economic forces, has restricted effectiveness.

Similar issues emerge in Chapter 9, the subject of which is Kenya's District Focus Policy for Rural Development. Marcel Rutten traces the evolution of development planning from its post-war British colonial roots through successive centrally controlled national development plans that embrace increasingly explicit spatial components and attempt more horizontally oriented initiatives. Disappointing results and the great changes to both the domestic and external environment for planning in the early 1980s provided the foundations for the District Focus Policy. One of its key objectives has been to overcome the structural contradiction between vertical integration of sectoral ministries at all spatial scales of a highly centralized state in which functional regional planning remains paramount, and horizontal integration required for district-level territorial planning. General discussion of the policy is followed by a detailed examination of the experience in Kajiado District. Rutten concludes that, although considerable progress has been made, many further obstacles to flexible, participative and responsive district-level planning remain.

In the following chapter, Allert van den Ham and Ton van Naerssen's study of lower level planning in Indonesia shows that such problems are by no means peculiar to Africa. In the very different environment of a densely populated archipelago state, the long traditions of highly centralized and authoritarian government are proving difficult to overcome. The mere creation of decentralized planning structures does not guarantee operational coherence, effective local participation or a more equitable distribution of resources among social groups. Van den Ham and van Naerssen illustrate their arguments with a case study of the implementation of the Strategic Development Framework in Sukabumi District in western Java in the light of changing conditions during the 1980s.

In Chapter 11, Ruud Jansen and Paul van Hoof provide a rather more optimistic assessment of recent regional planning initiatives for rural development in Botswana, one of the few multiparty parliamentary democracies and economically prosperous countries in the Third World. Notwithstanding the fact that a ruling élite exercises a high degree of control over the organs and resources of the state, and similar problems exist with district-level planning as in Kenya, considerable progress has been made towards the creation of viable and replicable rural planning at subdistrict level during the 1980s. This underlines one of this book's central themes, namely that implementation and evaluation of regional planning are meaningful only in relation to specific state forms and social relations. Jansen and van Hoof analyse the Communal First Development Area strategy by means of case studies of two contrasting districts, and also examine the improved land-use planning experiment in Southern District.

Botswana's prosperity contrasts starkly with the economic collapse experienced by Ghana since the early post-independence period. In Chapter 12, Bill Gould surveys that country's plight, the bitter pill of structural adjustment the IMF forced the Rawlings regime to swallow, and the nature of decentralized planning that has emerged as part of the new order and the Economic Recovery Plan. Although urban-rural income differentials declined dramatically as a result of the crisis, population circulation and migration continue. Marked regional discrepancies persist because previous policy favoured commercial agricultural production, concentrated largely in already developed areas, over remoter regions dominated by subsistence cultivation. With reference particularly to educational planning, Gould examines the nature of the newly created district structures and concomitant reorganization of ministerial decision-making and implementing mechanisms. Given their recency, evaluation would be premature, but there are considerable potential advantages. Crucially, progress will for a time be dependent on continued external resource inflows. Perhaps, structural adjustment with a human face and making a significant contribution to social improvement will yet be possible.

Jan Sterkenburg's discussion, in Chapter 13, of data needs and data systems for regional planning in Kenya and Sri Lanka, highlights some of the very practical problems to be overcome in implementation of appropriate and flexible subnational planning. Notwithstanding the contrasting characteristics of the state, types of decentralized planning being implemented and prevailing economic and political conditions in his two case studies, many common obstacles arise. These centre on data availability and collection, the lack of uniformity and co-ordination between different sectoral ministries regarding their basic spatial units of organization, frequent changes to such units, the absence of effective monitoring procedures within planning and personnel problems that restrict the potential advantages of computerization.

Finally, in the concluding chapter, Carole Rakodi and David Simon draw together some of the salient features to emerge from the various studies, and highlight the insights they have provided for regional planning. The combination of single-country foci and comparative studies has proved very useful in casting new light on recent planning practice across the spectrum of state forms and national conditions in the Third World. Recent conceptual advances in our understanding of the dynamic interdependence of society and space are now informing analyses of current planning practice. This is an essential step towards the incorporation of the lessons into such practice itself. Were this to happen, at least in part, thus contributing to the empowerment of the poor wherever in space they find themselves, regional planning and development really would come of age.

Reference

Abdallah, I. S. (1978) Heterogeneity and differentiation: the end for the Third World?, *Development Dialogue*, 1978: 2, pp. 3–21.

PART I: ANALYTICAL FRAMEWORKS

1

THE QUESTION OF REGIONS
David Simon

The contemporary Third World context

The 1980s have been a period of crisis and change in the global economy and none have felt the impact more acutely than the billion or more impoverished people across the Third World. For most of them, the prospects of improving their miserable conditions of existence appear no better – and sometimes worse – today than a decade ago. At a time when rapid technological progress, the development of crop varieties suited to conditions in many food-deficit areas and greatly increased awareness of the need to pursue environmentally sustainable development paths should be having a positive impact, they continue to suffer the debilitating effects of drought, famine, war, political repression, exploitation and debt.

The optimism and dreams of the heady 1960s and early 1970s have faded; yet the nightmares of poverty, underdevelopment and grossly unequal income distribution remain. Newly independent states, supported by international financial institutions and agencies, implemented costly capital investment programmes designed to promote modern, industrial development in the Third World. The failure of the majority to secure an industrial future has been manifest; yet even most of the handful of 'newly industrializing countries' hailed as success stories are today suffering acute structural problems, indebtedness and widespread environmental degradation.

It is not only capitalist development strategies that have proved less of a panacea than anticipated by their proponents. With the possible exception of Cuba, and in some respects also China, post-revolutionary states seeking socialist paths to development have encountered formidable obstacles, despite often dramatic progress with social programmes. Over the past decade they have been forced to abandon socialist orthodoxy for more 'pragmatic' links with foreign sources of capital investment, both corporations and those distrusted agents of international capitalism, the IMF and World Bank.

Across the Third World, then, the 1980s have been characterized by the debt crisis and structural adjustment lending on rigid conditions imposed by the IMF and other major donors. Inevitably, the extent of crisis varies between countries and continental regions. Notwithstanding the dramatic problems faced by the major Latin American debtors and much of Africa (e.g. Onimode, 1988; Save the Children Fund and Overseas Development Institute, 1988), there are some notable exceptions. Many Asian countries have made significant progress over the last 10–15 years, and their current prospects are commensurately brighter. All too often, however, primary concern has perforce shifted from development promotion to the necessity of ensuring short term survival. The financial centres of global capitalism in the 'North' are thus arguably in a position of strength unparalleled since decolonization, to dictate the direction and terms of 'development' in many parts of the Third World. The tone was set early in the decade by the Berg Report (World Bank, 1981), which saw 'accelerated growth' as the way forward. This reflected the reassertion of earlier conservative positions that growth ('efficiency') should take primacy over redistribution ('equity') and that state intervention should be minimized in favour of market mechanisms. Partly in consequence, and rather more promisingly, many grassroots movements have taken the lead in organizing more sustainable development strategies at the local level, often with assistance from foreign non-governmental organizations (NGOs). Nevertheless, rural surplus producers have generally benefited from the higher producer prices introduced under structural adjustment or, as in Zimbabwe, as part of 'socialist' rural development (Rakodi, this volume). Furthermore, evidence is now emerging that adjustment programmes have begun to devote greater attention to social, as opposed to purely economic, spheres (Gould, this volume).

Regional policy has come to form a significant element of development planning since the 1960s, although defined and implemented in different ways by contending schools of thought. Like other aspects of formal planning, regional planning evolved in North America and Europe and was exported to the Third World as part of the recipe for attaining modernization and development. Although clearly representing rather more of a continuum than a simple dichotomy, the respective strategies can be divided into broadly 'top down' and 'bottom up' categories according to their predominant emphases. Each contains examples from both the neoclassical and various political-economy traditions. Perspectives and the prescriptions based on them have changed in the light of earlier failures. It is commonly accepted that regional development has, to date, had only very limited success in achieving its objectives. The last few years have consequently witnessed vigorous debate among researchers and policy makers alike, as to whether only the efficacy of the tools or the entire conception of space underlying 'spatial policy' has been at fault.

Yet, the neoliberal development policies currently being promoted as the new orthodoxy by the World Bank, regional development banks, USAID and similar agencies contain only a watered-down regional component, styled 'decentralization', which takes no account of these important debates.

Both the regional terminology and strategies purportedly based thereon have come to mean such different things to different people that the whole question of regions and regional policy warrants rethinking. Are planning regions ultimately a useful concept, and is regional policy an effective means of overcoming spatially uneven development? 'The answer', according to Bob Dylan's famous refrain, 'is blowing in the wind.' For, even acceptance of the substance of many recent critiques invites serious problems of operationalizing alternative strategies in the current planning environment faced by Third World countries. This book, like the symposium out of which it has developed, seeks to take stock at this important juncture by reviewing the debate, examining recent experience with different types of policy in diverse Third World settings and highlighting directions for future progress.

Changing perspectives on regional development

This section does not attempt to provide a detailed overview of the various contributions to the field; substantive recent surveys can be found in Stöhr and Taylor (1981), Weaver (1981), Moulaert and Salinas (1983), Forbes (1984), Gore (1984), Dewar, Todes and Watson (1986) and Hönsch, Lavrov and Sdasjuk (1986). A rather different, curiously structured and unnecessarily dense account is given in Riddell's (1985) text, apparently aimed at practitioners. Instead, I merely highlight some of the key ideas of successive perspectives necessary to my argument and to provide a context for the remainder of the book. Little is said about the evolution of spatial policy in socialist countries of the Third World, since this forms the subject of Chapter 2.

Conventional 'top down' policies

'Scientific' regional planning formed an important element in industry-led development strategies. It is derived from modernization theory and the emerging discipline of regional science pioneered by Walter Isard (1956a, 1956b, 1960, 1975), – both of which are underpinned by neoclassical economics. Regional planning ostensibly provided the means whereby economic and social development could be attained even in remote and peripheral spaces. Deriving inspiration from seminal work by Myrdal (1957), this basically 'liberal' school of thought held that state action was necessary in order to overcome centripetal or backwash effects that drain the periphery of resources and relatively skilled labour towards the core region(s). By promoting spread or trickle-down effects, regional inqualities could be reduced and eventually eliminated by the attainment of a relatively homogeneous development surface across the national territory. The approach is perhaps best exemplified by John Friedmann's (1966) four-stage model of spatial development, itself strongly influenced by Rostow's (1960) concept of economic growth as a linear sequence of stages.

This interventionist paradigm stood in opposition to the *laissez-faire* position of more conservative neoclassical economists (e.g. Hirschman, 1958), who argued that centralization of urban industrial growth should be permitted so as to obtain the advantages of economies of scale. When these were outweighed by diseconomies, spontaneous dispersal of economic activity would occur. In other words, market forces should be allowed to take their own course; market imperfections did not warrant intervention. They held that, particularly in Third World countries characterized by scarce skills and development resources, attempts to force diffusion prematurely would be detrimental to economic growth, the engine of development. Riddell (1985, p. 45) reveals great confusion on this score, calling diffusion theory 'radical' and equating it with public intervention to promote the spread of development.

Many geographers and planners worked hard to promote regional planning as part of the conventional wisdom, and felt vindicated as more and more Third World countries incorporated explicitly spatial dimensions into their Five Year Plans during the 1970s. The prescriptions almost invariably involved the designation of a set of growth centres at one or more levels of the urban hierarchy in each demarcated region, coupled with a package of incentives to attract the desired investment. Most initial strategies focused on industrial expansion in a small number of large growth centres. Some countries also introduced disincentives in order to discourage further expansion of existing metropolitan cores.

If energetically applied, these measures were felt capable of overcoming centralized political and economic forces vested in the metropole, thus diffusing 'development' through the national space. In the light of experience, there have been many attempts to improve the tools for promoting decentralization and diffusion. These include a second, more sophisticated, generation of growth centre strategies integrated across sectors and (at least in theory) directly related to overall national development policies (Appalraju and Safier, 1976; Ternent, 1976). A number of variants of such explicitly 'spatial policy' now exist, styled 'national urban development strategies', 'national urban policies' and the like. Whereas initial strategies focused on the promotion of propulsive growth in a few large cities, a principal objective of more recent ones has been to curb continued metropolitan growth in favour of small and intermediate urban centres, sometimes as an element of rural development. Among the suggested potential contributions to spatial equity and national development such lower-order centres can fulfil under the right conditions (e.g. Rondinelli and Ruddle, 1978; Renaud, 1981, Mathur, 1982; Rondinelli, 1983a, 1983b, 1983c) are

- improved access to services by their residents and those of surrounding rural areas;
- better provision of shelter and physical infrastructure to improve mobility and access to markets;
- promotion of agricultural diversification and the production of surpluses;
- stimulation of small-scale and labour-intensive industry;
- better utilization of local resources, and
- retention of population who would otherwise migrate to large centres.

The costs of providing shelter, infrastructure and 'social overhead capital' such as education and health facilities, are often claimed to be lower in intermediate cities outside metropolitan areas, although Richardson's (1987a, pp. 221–30) recent comparative study shows that lower per-capita costs are partially offset by increasing costs with distance from the core region.

Part of the impetus to the proliferation of small- and intermediate-centre strategies was almost certainly provided by the 'redistribution with growth' school of thought promoted by the World Bank and ILO, in particular, during the early and mid-1970s (Chenery et al., 1974). As its name suggests, the object was to reduce societal inequality by favouring the poor through investment in appropriate sectors as the economy grew. Several possible mechanisms existed, centred on growth maximization, redistribution of investment, of income or of existing assets respectively, but overall economic efficiency and growth were not to be sacrificed. Although rural development was stressed, this was very much still a top down approach (see Dewar, Todes and Watson, 1986, pp. 73–85).

Recent empirical research has revealed that small- and intermediate-centre strategies have generally been inadequately focused and poorly implemented. This is due in part to the persistence of the master-plan mentality in Third World state bureaucracies, whereby the emphasis has been on physical planning to the neglect of economic and social relations. Blanket measures are also commonly applied across the set of 'target' centres without taking due account of specific regional contexts and local conditions in each case. External advice and assistance often still draw on inappropriate Western concepts and experience, while few expatriate consultants spend long enough in the field to familiarize themselves with individual centres. Crucially also, the implicit effect of many existing aspatial government policies relating to the economy and pricing of agricultural and basic commodities, tax structures, sectoral investment priorities and centralized political control at the expense of local authorities, continue to favour metropolitan core regions, thus undermining efforts to promote smaller centres (Renaud, 1981; Richardson, 1987a, 1987b; Blitzer et al., 1988; Hardoy and Satterthwaite, 1986, 1988; Hinderink and Titus, 1988; Choguill, 1989).

Nevertheless, various related genres of decentralization policy, geared to facilitating the wider distribution of infrastructure and services across the national space, have become the new orthodoxy of international agencies and donors. Most rely on similar conceptual underpinnings to growth centre and integrated rural development strategies, are totally eclectic or are atheoretical, deriving their inspiration from the neoliberal obsession with debureaucratization, deregulation, privatization, individual freedom and 'choice'. A good recent example is provided by Rondinelli et al. (1989), who ignore many possible reasons for, and advantages of, centralization in the Third World; deal with space and infrastructure in isolation from the territorial expression of power; and overlook the stark fact that the rhetoric of decentralization is frequently accompanied in reality by further centralization of power (see also Slater, 1989). For a contemporary example of the last mentioned, we need look no further than Thatcher's Britain.

It is worth noting two further points about small and intermediate urban centre strategies. First, despite their focus on urban dispersal, decentralization or deconcentration, such policies remain top down, being articulated and implemented – to the extent that they actually are – from the political centre. Second, this objective has not been the sole preserve of Western countries and basically procapitalist Third World states. Many socialist states in Eastern Europe and the

Third World have implemented similar strategies, albeit legitimized in a different fashion and sometimes better articulated with other elements of territorially based development (see Slater, 1978, 1982, 1987; Susman, 1987; Sidaway and Simon, this volume).

From the foregoing analysis it is evident that top down planning, or development from above, has seldom been successful even where sufficient time has elapsed to permit evaluation of the overall impact of programmes initiated twenty to thirty years ago. In empirical terms, therefore, the thrust of Gilbert and Goodman's (1976) critique is still valid today: modern 'development' is largely restricted to urban centres and their immediate surroundings, while inter- and intraregional disparities in material wellbeing have generally not declined. 'Place poverty' and 'people poverty' are often still being inadequately distinguished or are even regarded as synonymous. Regional inequalities have generally increased, with the poorest regions worse off than before (Krebs, 1982). If rural development is the object, direct investment in small rural centres would be more appropriate as one element of strategy (Wong and Saigol, 1984).

Growth centre strategies are a spatial derivative of the aspatial growth pole concept proposed by the French economist, Perroux (1950). He suggested that concentrating investment in dominant or propulsive industries could achieve dynamic growth through the stimulation of backward and forward linkages so as to maximize the investment multiplier. Translation of growth poles into geographical growth centres owes much to the work of Boudeville (1966), Friedmann (1966), Hansen (1967, 1972, 1981) and Darwent (1969), among others.

The fact that such strategies are still actively promoted in various guises today, despite their poor track record, attests to their conceptual attractiveness. At least three reasons for this have been distinguished: the logical appeal of the concept of concentrating rather than dispersing investment so as to maximize its impact; the observed experience in advanced capitalist countries where urban and industrial growth have been synonymous and have underpinned development; and the technocratic authority lent by the concept's origin in the popularly recognized discipline of economics (Friedmann and Weaver, 1979, p. 125; Forbes, 1984, p. 116). To this I would add a fourth, namely the concept's apparent simplicity. Yet this simplicity has provided perhaps its greatest flaw, given the absence of any rules or prescription for determining the appropriate number and sizes of growth centres, their distance apart or relationship to designated regions. Much research and debate has gone into optimum city size and related questions of settlement policy (notably the work of Richardson, 1972, 1973a, 1973b, 1976, 1981), but without the achievement of any consensus. Given the wide range of actual conditions in Third World countries, this is probably inevitable. Gore (1984) provides an important fifth reason for the persistence of such strategies, namely their political utility to the state. This will be discussed in greater detail in the final section below.

A third major strand of non-radical writing on regional development emerged during the 1970s in response to growing evidence that growth centre strategies were not succeeding. Instead, capitalist economic growth was seen as exacerbating interregional inequalities through forces of circular and cumulative causation. The result could eventually be instability. John Friedmann's (1972, 1973) theory of polarized development was indicative of these growing perceptions. Friedmann's writings over the last twenty years provide a fascinating chronicle of changing paradigms and praxis, from optimism over modernization to concern at growing regional imbalances and, as we shall see below, to advocacy of apparently more radical alternatives such as agropolitan development and territorially based approaches.

Finally, in this section, it is necessary to highlight a fundamental flaw in the conceptualization of space evident in the work of conventional 'spatial development' schools. Charles Gore's (1984) sustained and cogent critique of existing regional development theories echoes earlier radical analysts in arguing that these theories all separate space unacceptably from other social, economic and political aspects of life (see also Coraggio, 1983; Markusen, 1983). Moreover, by so doing, space is implicitly and incorrectly regarded as having causal powers. This 'incomplete relational concept of space' undermines not only the theory but also policies derived from it. The gulf between rhetoric and reality is often great (Slater, 1989). In a more general context, Sack (1974, 1980, 1981) has advanced similar arguments on the need for a fuller relational conception of space in analysing the territorial bases of power.

'Top down' political economy perspectives

Conventional perspectives view capitalism, with its focus on profit-oriented individual enterprise, as the most desirable economic system and essentially benign, even if manipulation by public intervention is necessary to maximize the common good. By contrast, political economy ('radical', i.e. Marxist, neo-Marxist, structural Marxist) schools of thought regard capitalism as inherently exploitative, generating class conflict and spatial inequality. More equitable development can only be achieved by a transition to socialism. The respective orientations, assumptions and dialectics are thus fundamentally different (Coraggio, 1983; Moulaert, 1983).

There is a significant body of writing on various aspects of underdevelopment and regional devolution and cultural issues in the First World (Weaver, 1981; Forbes, 1984). Much of the relevant literature on the Third World has comprised critiques of existing practice and the refinement of alternative theoretical frameworks. There is remarkably little material suggesting concrete strategies based thereon or, indeed, evaluating those which have been attempted in socialist states (see Chapter 2). Fundamentally, Marxists are concerned with modes of production and their attendant relations of production and outcomes, of which space is one dimension. But, as Forbes (1984, pp. 118–19) indicates, there has been debate among radical geographers about how far the spatial aspects of capitalism can be considered without abstracting space from its relational context and implicitly according it autonomy and causal efficacy. Some analyses have left themselves open to precisely this criticism, e.g. by Gore (1984, pp. 197–8), as discussed below. Spatial separatism is not a flaw solely of conventional spatial development theory.

The most important strand within the political economy literature, originating in Latin America, is concerned with dependent development or underdevelopment.

The stimulus was provided by Andre Gunder Frank's (1967, 1969, 1978) neo-Marxist formulation of what has come to be known as dependency theory, partially inspired by Baran's (1957) work on the US economy. Frank's thesis was that the global system of international capitalism was dominated by an alliance of transnational corporations and states based in the North, which dictated the nature and extent of Third World economic development according to their own interests. There was no significant local autonomy or real development, since local élites were maintained by, and beholden to, foreign capitalist interests, while only economic and social development functional to these interests was permitted.

In other words, development in the First World core was predicated on underdevelopment in the Third World periphery. Hence capitalism as found in the Third World was a dependent and distorted form of metropolitan capitalism. This was a gloomy and deterministic analysis: little could be done to remedy the position short of severing links with the world economy. There was clearly no room for spatial policy or any other form of local development initiative. Similar analyses of the African experience soon appeared, associated with Samir Amin (1974, 1976) and Walter Rodney (1972), among others.

The dependency conception was thus the polar opposite of conventional theory, in terms of which foreign capital investment provided the primary vehicle for development in Third World countries. It is also distinct from the essentially structuralist writings of Raul Prebisch and colleagues at UNECLA, advocating import substitution and other measures to reduce external dependence.

Frank's ideas were elaborated in spatial terms by Osvaldo Sunkel (1973) and José Coraggio (1975), focusing respectively on the roles of transnational corporations and growth centres in exploitatively concentrating available skills and resources at the expense of host-country economies, and in creating new points of surplus extraction in peripheral areas, thus promoting underdevelopment in dependent space economies. To quote Weaver (1981, p. 84), beneath its accusative tone, the theory's central argument is that 'Functional economic power, removed from the control of territorial authority, can only exacerbate the social and geographical inequities inherent in polarized development'. By implication, no spatial development policy based on capitalism could have a positive impact under such conditions.

In a similar vein, David Slater's early work (e.g. 1975) highlighted how precolonial spatial organization in countries like Peru and Tanzania had been totally disrupted and transformed by the dictates of colonial capitalism (see also Salinas, 1983). Post-colonial territorial integration for the benefit of the indigenous population would require the reorientation of colonial transport, economic, social and political networks, which were geared to domination and resource exploitation instead of local development.

Gore's (1984, pp. 197–8) important criticism centres on spatial applications of dependency theory. He points out that spatial dependency analysis is analogous to – but the opposite of – spatial diffusion theory, in that the transfer of surplus out of exploited regions actually occurs along networks of exchange relations. He argues that operationalization of the concept is problematic since, unlike diffusion patterns, these flows of surplus are difficult to trace; it is difficult to measure and define 'value' because market prices are not a reliable indicator; the conception of surplus may be separated (abstracted?) from the production process; and since the transfer of surplus carries no necessary implication that the recipient region(s) will grow faster. Like the spatial development strategies discussed earlier, this approach therefore suffers from an incomplete conception of space.

Dependency and underdevelopment theory in general, and the work of Frank and Amin in particular, have been severely criticized on both empirical and theoretical grounds (e.g. Smith, 1980; Schiffer, 1981) as being economistic, structurally deterministic to the exclusion of human agency or local autonomy and neglectful of class analysis – the centrepiece of Marxist political economy. Many of the original authors now admit to these shortcomings and have since explored more sophisticated forms of analysis of regional problems, embodying critical political theory. Incorporation of a theory of the state has been particularly important (Coraggio, 1983; Markusen, 1983; Slater, 1988). Another such approach has been the study of spatial divisions of labour, a theme popularized by Massey (1984). Although theoretically informed analysis is a necessary precursor to action, there has seldom been a direct and practical alternative policy dimension to such work.

One potentially fruitful avenue explored during the 1980s has been an attempt to integrate structure and human agency within political economy. Taking inspiration from the work of Anthony Giddens (1981), this structurationist paradigm opposes functionalist theoretical explanations and seeks to avoid both structural determinism and voluntarism. The need to include space, time and a theory of practical social action, all of which are absent from most social theory, is also recognized (Forbes, 1984). Although no practical programme has emerged from this agenda, it has helped to refocus the attention of 'radicals' on people as both the subject and object of development, taking important structural constraints and consequences into account.

'Bottom up' strategies

Whatever the precise mechanisms and degree of state involvement, all permutations of 'development from above' are predicated on the supposed outward and downward diffusion or extension into the periphery of innovations, development or exploitative relations of production from core regions of urban industrial concentration and capital accumulation. The strategies are conceived, organized and implemented from the centres of political and economic power, often with little (if any) regard for the views and interests of supposed 'target' areas and groups.

Growing disillusionment with this approach has centred on its assumptions and methods, and its general failure to achieve results commensurate with expectations or the resources allocated. Further problems have been regional resistance to the continued central domination and 'cultural imperialism' usually thereby implied, the general concern in reality with economic growth ahead of development and the continued poverty and powerlessness of the majority of Third World inhabitants, both urban and rural. During the 1970s, many social scientists therefore searched for meaningful alternatives that would focus first and foremost on the poor, those who were being bypassed or further exploited and impoverished by existing patterns and relations of production. Historically, there have, however, been periodic swings in many long-established countries between policies discernible as top down and bottom up, in accordance with changing cultural, political and economic conditions (Stöhr, 1981, pp. 47–61).

Development from below embodies a range of supposedly bottom up strategies that reflect the differing perspectives of the authors, but all profess to being people centred in a manner diametrically opposite to top down policies. Hence, many of these approaches, articulated by writers such as Paul Streeten (Streeten and Burki, 1978; Streeten, 1980), have come to be known by the umbrella term of basic needs strategies (see also Dewar, Todes and Watson, 1986). As the name suggests, the object is to provide the poor with the minimum levels of nutrition, safe drinking water, shelter, hygiene, health, literacy and so forth to enable them not merely to survive but to improve their own circumstances. This involves harnessing local resources, potential and initiative for the benefit of the people themselves. Put simply, it is helping poor people to help themselves.

Such thinking has increasingly been adopted in the work, for example, of the more progressive NGOs involved in international aid and development. Projects and schemes are generally small scale, employ appropriate technologies and maximize utilization of locally available inputs. The degree of outside advice, assistance and aid varies, but success depends crucially on the local beneficiaries being actively involved and exercising a significant degree of control from the beginning. Ultimately, of course, they should be trained to take full control. Moreover, in the important ongoing debate over how to promote environmentally sustainable development, the crucial role of such small scale local community action by hitherto marginalized and exploited groups has recently been recognized (e.g. Redclift, 1987; Simon, 1989).

For some writers and organizations, the basic needs strategy is an end in itself, but other theorists and planners envisage it as a step towards national development whereby the social benefits and surplus produced eventually contribute to elimination of the periphery. Inevitably, perhaps, popularization and inclusion of basic needs in the development jargon have led to devaluation of its essence in many guises. In a trenchant recent rebuttal of the growing disillusionment over the apparent failure of many basic needs strategies, Wisner (1988) argues that most governments and multilateral development agencies are guilty of just such devaluation. Behind the rhetoric of helping the poor to improve their circumstances is the reality of continued top down activity, with the poor reduced to being passive recipients of often inappropriate goods and services supplied individually rather than on an integrated basis. This shopping list approach, characteristic of conventional development paradigms, has reduced basic needs to a 'weak' form in which the poor are still marginal. By contrast, 'strong' basic needs requires that the poor be central. In essence, therefore, the poor must be *empowered*. Therein lies the reason for the failure of governments and agencies to implement 'strong' versions of the approach: it would involve fundamental changes to the status quo and the distribution of power within society. It is not functional to governments and their external supporters. One important strand of 'strong' or radical basic needs thinking has emanated from the Dag Hammarskjold Centre in Sweden over the last fifteen years. Originally labelled 'Another Development', its most recent formulation is 'Human Scale Development' (Max-Neef *et al.*, 1989). However, since progressive NGO operations are insignificant compared with those of governments and multilateral agencies (including most within the UN system), Wisner's claims that 'strong' basic needs approaches have seldom been given a chance, and that it is 'weak' basic needs strategies that have failed, have much justification.

In this context, it is worth making the point that even concerted rural development strategies purporting to be bottom up are often not that. In one of the best-known attempts at socialist development, Tanzania adopted an inward-looking programme, loosening its dependence on unequal external trade and promoting a comprehensive, rurally oriented self-reliance programme of which the *ujamaa* villagization and collectivization schemes formed the centrepiece. Dissatisfied with the slow rate of voluntary action, the state became coercive, resettling large numbers of villagers and revealing the programme as a highly centralized state strategy aimed, as in so many other Third World countries, at capturing the peasantry and incorporating them into the dominant mode of production (e.g. von Freyhold, 1979; Hyden, 1980). In addition to the loss of legitimacy engendered by coercion, another significant reason for the failure of *ujamaa* was – importantly for present purposes – the absence of an appropriate spatial component to the strategy. A single blueprint was implemented nationally, without regard to regional differences in environmental conditions, culture and social organization (Luttrell, 1971; also Rakodi, this volume). There have been various attempts to elaborate basic needs in a spatial context, of which the best known is agropolitan development (Friedmann and Douglass, 1978; Friedmann and Weaver, 1979; Friedmann, 1980) and the related necessity for selective spatial closure (e.g. Stöhr and Tödtling, 1978).

It is important to distinguish between planning focused on *functional* integration of the space economy, whereby the resources of a region are exploited purely for their potential contribution to the wider economy, and *territorial* integrity, in terms of which a region's resources are developed for the benefit of the local population (Friedmann and Weaver, 1979; Stöhr, 1981; Weaver, 1981). By definition, therefore, basic needs strategies, such as agropolitan development, fall into the latter category. Although functional regional planning is generally top down and territorial regional planning bottom up, they should not be regarded as synonymous or the two approaches as being mutually exclusive since cross-cutting examples are conceivable.

Agropolitan development, as elaborated in successive versions for Asia and then parts of Africa by Friedmann and his colleagues (Friedmann and Douglass, 1978; Friedmann and Weaver, 1979; Friedmann, 1980), evolved from being 'the spatial correlate of a strategy of accelerated rural development' to being 'a basic needs strategy of territorial development' (see also Gore, 1984, pp. 164–9). Basically, it amounts to an integrated, locally controlled rural development strategy, in which each agropolitan district is centred on a town of 10,000–25,000 people and forms the basic unit of local government. District boundaries should encompass commuting radii of not more than 5–10 km, and coincide with 'local communities of interest'. The envisaged strategy is inward looking, involving a high priority on local self-sufficiency through development based primarily on agriculture. However, ancillary production of wage goods through light industry, replacement of generalized and unlimited 'wants' as defined in conventional economic growth theory with specific limited needs, conservation of local ecosystems as part of development based heavily on local skills and knowledge, reduction of existing inequalities, and provision of social services and infrastructure, are also important.

In recognition that aspects of both 'territory' and 'function' are necessary for overall development, total isolationism is not advocated. However, a strategy of selective spatial closure at the appropriate territorial scale of integration is necessary in order to ensure that the benefits are retained by local people. Communal ownership of productive wealth (presumably the means of production and output), and equalization of access to the bases for accumulating social power, are required to ensure that the whole community benefits. Large scale, capital intensive industries and facilities will also be necessary, but to avoid competition with agropolitan districts, they should be concentrated in enclaves in a strategy of 'planned industrial dualism'.

In a related article on urban poverty, Friedmann (1979) elaborates the bases for accumulation of social power as comprising the following elements: appropriate knowledge and skills, productive assets, financial resources, social and political organizations and social networks. Although very few case studies of agropolitan development have appeared in the literature, the emphasis on the empowerment of the poor qualifies the approach as a 'strong' basic needs strategy, in Wisner's (1988) terminology, and is certain to suffer the same obstacles to implementation that he identifies.

More generally, territorial regional planning can be described as constituting a method of ensuring locally appropriate development through maximum mobilization of endogenous material, human and institutional resources. Key features embrace (Friedmann and Weaver, 1979; Stöhr, 1981; Weaver, 1981):

- enabling broad access to land and other key forces of production;
- evolution or revival of territorially organized structures for equitable communal decision-making;
- greater powers of self-determination for local communities and areas;
- choice of regionally adequate and appropriate technology;
- giving priority to the satisfaction of local basic needs;
- enhancing the terms of trade for agricultural and other products of the periphery;
- local control over external assistance in cases where local resources and skills are inadequate in order to ensure compatibility with local objectives;
- restricting the production of exports to a level where the proceeds contribute to a broad improvement of local quality of life;
- restructuring urban and transport systems to improve and equalize access to them from all parts of the country;
- improving intraregional and especially intrarural transport and communications; and
- engendering egalitarian social structures and a collective consciousness.

Implementation of this approach invariably requires selective spatial or territorial closure, to help retain locally produced surplus and surplus value for local benefit, and to prevent the undermining of local initiatives by external competition, especially from the large scale sector benefiting from economies of scale. Interaction can thus take place on the terms of the sheltered territorial unit. No consensus exists, however, as to whether closure should be a permanent or only a temporary arrangement. In essence, therefore, this is a spatial application of the neoclassical economists' so-called infant-industry argument. Specific imports are restricted through quotas or tariff barriers to assist a newly established local industry become viable and competitive. However, unlike the territorially based strategies being discussed here, infant-industry theory states explicitly that the barriers should be removed in due course in order to ensure continued competitiveness and economic efficiency of the industry. Nevertheless, this frequently does not occur in practice. Funnell's (1986) interesting case study of selective spatial closure in operation shows clearly that social and economic inequalities may in fact be exacerbated, despite (or because of) retention of locally produced surplus value. Development policies therefore cannot be solely spatial in nature.

In an important critique of territorial regional planning, Gore (1984, p. 165) claims that Friedmann and Weaver are overly idealistic in their plea for 'the recovery of territorial life', investing it with utopian qualities. More fundamentally, their conception of territorial units is problematic because the notion of space used is derived from biology and is inadequate for understanding and planning the social world (p. 229). This problem has three aspects (pp. 230–1):

1. Territorial units at all scales (national, regional, district) are treated as organisms. However, to sustain the analogy, the country would have to be the organism, and its various components the organs, not supposedly autonomous organisms.
2. Regions are not objective organisms with a life of their own, but are subjectively defined by people and social relations.
3. The belief that the restoration of territorial integrity will release a force of 'wilful action' through which the local population will develop themselves, derives from the long-discredited biological notion of vitalism.

How important the organismic analogy in itself is to territorial regional planning, and hence how far we ought to be concerned with this aspect of Gore's critique, is debatable. However, it is an element in a wider critique that goes on to suggest that territorial regional planning is not as complete a paradigm shift from functional regional planning as its proponents argue. For the assumption of functionalism is no less important to this approach, and both paradigms (necessarily) encompass the significance of space in determining human activity:

> Indeed, *territorial regional planning reflects adherence to the spatial separatist theme in a new guise.* Just as in spatial policy, elements of the spatial structure, or spatial relationships, or 'locational factors' are taken to be causes, so, in territorial regional planning, the region is given causal efficacy. … when a region is treated as an 'organic whole', the behaviour of its elements is attributed to *the intrinsic nature of the region itself.* Indeed any attempt to explain the behaviour of individual elements by reference to social and economic processes is misplaced. For 'the whole is more than the sum

of its parts' ... the outcome of spatial policy depends upon underlying social and economic processes. But territorial regional planning, which aims to arrange the phenomena of social life into 'organic wholes', *makes these processes dependent upon the nature of the whole.* It is a neat inversion. But it must be rejected, unless one believes that the organic analogy provides a useful basis for understanding the changing world.

(*Ibid.* pp. 231–2)

Territorial regional planning, then, along with selective spatial closure and integrated rural development, all of which he labels 'neo-populist' strategies, suffers many of the same deficiencies as functional regional planning. Most fundamentally, consideration of space is separated from social, economic and political processes (i.e. 'spatial separatism') and is given causal significance.

Moudoud (1988) accuses Gore of being both unfair and misleading in his critique of these approaches, because the question of political and economic power is central to them. They are concerned essentially with a redistribution of power between all sociospatial entities at different scales, integral to which is local control over resources and their allocation. This response has some validity, since, as shown above, Friedmann has certainly been concerned essentially with the empowerment of the poor. However, Moudoud does not deal with the spatial organismic and utopian aspects of agropolitan development and territorial regional planning criticized by Gore. Also less dismissive than Gore is Boisier (1988), who argues that these new perspectives provide helpful insights, elements of which need incorporation into a radical but less utopian agenda for building regional societies.

Synthesis and prognosis: where do we go from here?

The foregoing analysis has shown most, if not all, spatial policies for regional development to be conceptually flawed in their treatment of space in relation to other facets of the political economy, and because the relationship between different geographical scales of analysis and action has frequently been inadequately articulated. Empirically, regional planning has generally proved rather less than successful, not least because sectoral (or aspatial) policies, which often conflict with spatial strategies, can have far more powerful spatial effects than explicitly spatial measures. Moreover, as Dewar, Todes and Watson, (1986, p. 157ff.) correctly point out, development theory and spatial planning have by and large been isolated from each other. The many theoretical paradigm changes have seldom brought about more than shifts in emphasis in spatial policies, which have remained remarkably consistent over time and, indeed, also between countries.

Why, then, do so many governments and agencies continue to promote regional development with such vigour? The answer, according to Gore (1984) and Slater (1988), is that such policies have great political utility in legitimizing the status quo. In Third World countries, the state is often highly centralized, beleaguered and unable to exercise full control over all its territory. Crudely, the rhetoric of decentralization or regional development is politically popular, particularly among those currently deprived of access to the bases for accumulating social power, to borrow Friedmann's (1979) term. Formulation of explicitly spatial policies backed up with resources is thus likely to relieve pressure on the state, while ingratiating it with international agencies and donor governments (some of which favour controlled decentralization) in an increasingly conservative 'neoliberal' global climate.

The theory, rhetoric and practice of regional policy can therefore only be understood if analysed as an integral facet of state policy within a theory of the state. Third World governments have two basic characteristics (Gore, 1984, p. 244):

1. The use of state power and finances to increase material production, by facilitating and supporting private capital accumulation and the establishment of state enterprises on a profit-making basis.
2. The attainment of legitimacy by claiming to represent the common interest and to be concerned with the national interest.

As Gore further argues, the benefits of such accumulation are, in reality, not equally distributed. The importance of the state means that it has a major influence over the distribution of benefits, yet its continued hold on power depends on the consent of politically powerful groups. This exposes the myth of complete commonality of interests between groups: 'The practical actions which are the substance of development policy may thus be understood as the outcome of a negotiation of conflicting powerful interests in society mediated through the institutional apparatus of the state' (*ibid.* p. 247). It is important, therefore, to appreciate that all such policies are not purely technical or 'scientific' but are also invested with an (often implicit) ideological and political content. For regional policies are 'chameleons', capable of achieving covert social objectives: 'To understand regional policies, one must take account of the fact that they are EXPLICITLY biased against spatially defined groups and IMPLICITLY biased in favour of socially defined groups' (*ibid.* p. 262). In view of the foregoing, we must analyse regional problems and action not only from the perspective of the state but also from that of the regions and, more specifically, regionally based groups. For, just as the state can manipulate territory and invoke spatial policies for its own ends, so too can disaffected groups and classes. The extent and intensity of regional tension or conflict in any country will obviously vary across both space and time as a reflection and articulation of perceived disadvantage, exploitation or domination. For the region is more than a cultural group, an economic system, or a purely political unit; it is essentially an arena within which social relations develop (Markusen, 1983). In a similar vein, Slater (1988, pp. 19–23) distinguishes four broad categories of regional problem, each articulated in a distinct manner:

1. Regionally based antagonisms within a power bloc, leading to the formation of regional blocs.
2. Popular demands for regional power and control, expressed through new regional social movements.
3. Indigenous expressions of ethnoregionalist identity, articulated through movements for indigenous autonomy [or even independence].

4. Challenges to state power by guerrilla organizations with specific territorial bases [even though unsurpation of state power may be the ultimate objective], waging armed struggle and insurgency.

It is not clear that these categories are necessarily mutually exclusive. For example, regionally based armed guerrilla struggle could, and frequently has, become the last resort of groups in the other three categories whose attempts to have their grievances redressed through other, non-violent modes of articulation, have been frustrated. Indigenous ethnoregional identity may first be expressed as demands for greater representation and a more equitable share of resources within a unitary state. However, this does not undermine Slater's basic thesis. Taking his analysis to its conclusion, a regional question or crisis may be defined in relation to territorially specific responses by the state, aimed at resolving a perceived threat to its political security or its own territorial jurisdiction and control.

This review has covered a wide field, attempting to highlight key features of competing and complementary approaches to regional planning and policy. The need for a broader, fully integrated, conception of space has been demonstrated. Theoretical analysis has certainly made significant strides over the last decade, driven by both the philosophical and empirical shortcomings of existing policies. However, neither Gore nor Slater has ventured beyond critical analysis into the arena of praxis, of developing alternative policies for state intervention appropriate to their diagnoses. That vital task is still waiting to be tackled. And in view of the fact that, notwithstanding changing emphases and some increasing sophistication, real-world practice seems generally to be lagging far behind, still largely fixated on space, this task has great urgency.

Finally, two very important questions emerge from the literature surveyed here. Have we now reached the point of denying the validity of spatial analysis and spatial policy – are regions redundant? And should all efforts at territorially based planning be abandoned? The implication of critiques by Gore and Slater, for example, may appear to suggest a resounding 'yes' in both cases. I am not so sure. Examine again the final step in Slater's analysis cited above. Is this really so radically different in essence from many conventional conceptions of the rationale for regional planning? Not at all, beyond the view of state action being negatively as opposed to positively motivated. Spatial or regional inequality has many dimensions: distribution of land, income, social welfare, standards of living, environmental quality and access to power. Any one or combination of these could be used by regionally based groups as a vehicle for articulating their demands. Any state response to this could be characterized as being negatively motivated. But, conversely, the state could use the same vehicle equally well with apparently positive motivation, either because it is genuinely concerned with improving conditions for its population or because it correctly anticipates future opposition from specific quarters should it fail to act. The answer is indeterminate, depending on the specific context. Either way, whatever action the state might take would almost certainly not be spatially neutral in its effect.

Some readers might accuse me of intolerable eclecticism at this juncture, on the grounds that increasing efforts to reach a *rapprochement* between the political economy and conventional traditions (as outlined by Dewar, Todes and Watson, 1986, pp. 118–23) are futile because they have diametrically opposed points of departure (e.g. Moulaert, 1983; Slater, 1988, pp. 3–7). But the point has been made: problems can be given whatever gloss is most convenient to the task at hand. While that slant does not alter the objective conditions of 'reality', it can influence or entrench powerfully the perceptions thereof held by opposing groups, thus affecting the course of political action.

Be that as it may, it seems to me, in summary, that we have now reached the following point. The state will act to preserve its overall control and territorial integrity in the face of perceived regionally based threats. The hows and wheres of this action depend on the nature of the state. Some interventions may be overtly geared to conditions in specific parts (regions) of its territory, but these are likely to affect certain groups within those regions adversely. Conversely, other groups elsewhere may in fact benefit. Groups are defined here in terms of their control over the means of production, or access to the bases for accumulating social power. Moreover, few state interventions are spatially neutral, having indirect spatial effects. Space is thus far from irrelevant. However, it must not be treated in isolation, but rather as an integral element of politico-economic systems. Society has little meaning without space.

This would seem to leave adequate scope for the formulation of multifaceted, integrated policies sensitive to different group and regional needs and embodying the most positive elements of both top down and bottom up planning. The guiding questions should be, what mechanisms, forces and structures determine the incidence of benefits and costs, and second, who benefits, how and where? Development efforts must be geared first and foremost to the needs of the underclasses, however delineated. And empowerment of the poor is, by definition, a radical undertaking. For precisely this reason, few states can be trusted to promote their interests with vigour.

Hence grassroots action to secure adequate levels of living with a maximum degree of self-reliance and control is ultimately the best way forward. It should be promoted wherever possible with appropriate external assistance. As Boisier (1988, p. 54) concludes, 'The building of a regional society can only be accomplished with and by the regional community, even if this community is, in the beginning, incipient and ill-defined. Outside aid, which is normally needed at the outset as an inductive mechanism, should be halted as soon as possible.' This is the essence of 'strong' basic needs and 'Human Scale Development' – a potentially fruitful avenue for development planning on which new light is cast by several of the case studies in this volume.

References

Amin, S. (1974) *Accumulation on a World Scale*, Harvester, Hassock.

Amin, S. (1976) *Unequal Development*, Harvester, Hassock.

Amin, S. (1977) Underdevelopment and dependence in black Africa: origins and contemporary forms, in J. Abu Lughod and R. Hay (eds.) *Third World Urbanization*, Maaroufa Press, Chicago, Ill., pp. 140–50.

Appalraju, J. and Safier, M. (1976) Growth centre strategies in less developed countries, in A. Gilbert (ed.), op. cit., pp. 143–67.

Baran, P. (1957) *The Political Economy of Growth*, Monthly Review Press, New York, NY (also published in 1973 by Penguin Books, Harmondsworth).

Blitzer, S., Davila, J., Hardoy, J. and Satterthwaite, D. (1988) *Outside the Large Cities: Annotated Bibliography on Small and Intermediate Urban Centres in the Third World*, IIED, London.

Boisier, S. (1988) Regions as the product of social construction, *CEPAL Review*, no. 35, pp. 41–56.

Boudeville, J. R. (1966) *Problems of Regional Economic Planning*, Edinburgh University Press.

Chenery, H. *et al.* (1974) *Redistribution with Growth*, Oxford University Press (for the World Bank).

Choguill, C. L. (1989) Small towns and development: a tale from two countries, *Urban Studies*, Vol. 26, no. 2, pp. 267–74.

Coraggio, J. L. (1975) Polarization, development and integration, in A. Kuklinski (ed.) *Regional Development and Planning: International Perspectives*, Sifthoff, Leiden.

Coraggio, J. L. (1983) Social spaceness and the concept of region, in F. Moulaert and P. W. Salinas (eds.), op. cit., pp. 21–31.

Darwent, D. F. (1969) Growth poles and growth centres in regional planning: a review, *Environment and Planning A*, Vol. 1, no. 1, pp. 5–31.

Dewar, D., Todes, A. and Watson, V. (1986) *Regional Development and Settlement Policy*, Allen & Unwin, London.

Forbes, D. K. (1984) *The Geography of Underdevelopment*, Croom Helm, Beckenham.

Frank, A. G. (1967) *Capitalism and Underdevelopment in Latin America*, Monthly Review Press, New York.

Frank, A. G. (1969) *Latin America: Underdevelopment or Revolution?*, Monthly Review Press, New York.

Frank, A. G. (1978) *Dependent Accumulation and Underdevelopment*, Macmillan, London.

Friedmann, J. (1966) *Regional Development Policy: A Case Study of Venezuela*, MIT Press, Cambridge, Mass.

Friedmann, J. (1972) A general theory of polarized development, in N. M. Hansen (ed.), op. cit., pp. 82–107.

Friedmann, J. (1973) *Urbanization, Planning and National Development*, Sage, London.

Friedmann, J. (1979) Urban poverty in Latin America: some theoretical considerations, *Development Dialogue*, 1979: 1, pp. 98–114.

Friedmann, J. (1980) The territorial approach to rural development in the People's Republic of Mozambique: six discussion papers, *International Journal of Urban and Regional Research*, Vol. 4, no. 1, pp. 97–115.

Friedmann, J. and Douglass, M. (1978) Agropolitan development: towards a new strategy for regional planning in Asia, in F. Lo and K. Salih (eds.) *Growth Pole Strategy and Regional Development Policy*, Pergamon Press, Oxford, pp. 163–92.

Friedmann, J. and Weaver, C. (1979) *Territory and Function: The Evolution of Regional Planning*, Edward Arnold, London.

Funnell, D. C. (1986) Selective spatial closure and the development of small scale irrigation in Swaziland, *Tijdschrift voor Economische en Sociale Geografie*, Vol. 77, no. 2, pp. 113–22.

Giddens, A. (1981) *A Contemporary Critique of Historical Materialism, Vol. 1: Power, Property and the State*, Macmillan, London.

Gilbert, A. G. (ed.) (1976) *Development Planning and Spatial Structure*, Wiley, Chichester.

Gilbert, A. and Goodman, D. E. (1976) Regional income disparities and economic development: a critique, in A. Gilbert (ed.), op. cit., pp. 113–41.

Gore, C. (1984) *Regions in Question: Space, Development and Regional Policy*, Methuen, London.

Hansen, N. M. (1967) Development pole theory in a regional context, *Kyklos*, Vol. 20, no. 3, pp. 709–25.

Hansen, N. M. (ed.) (1972) *Growth Centres in Regional Economic Development*, Free Press, New York, NY.

Hansen, N. M. (1981) Development from above: the centre down development paradigm, in W. B. Stöhr and D. R. F. Taylor (eds.), op. cit., pp. 15–38.

Hardoy, J. E. and Satterthwaite, D. (eds.) (1986) *Small and Intermediate Urban Centres: Their Role in National and Regional Development in the Third World*, Hodder & Stoughton, Sevenoaks.

Hardoy, J. E. and Satterthwaite, D. (1988) Small and intermediate urban centres in the third world: what role for government?, *Third World Planning Review*, Vol. 10, no. 1, pp. 5–26.

Hinderink, J. and Titus, M. J. (1988) Paradigms of regional development and the role of small centres, *Development and Change*, Vol. 19, no. 3., pp. 401–23.

Hirschman, A. O. (1958) *The Strategy of Economic Development*, Yale University Press, New Haven, Conn.

Hönsch, F., Lavrov, S. B. and Sdasjuk, G. V. (1986) *Bürgerliche Konzeptionen der Regionale Entwicklung*, Haack Gotha, Possneck/Mühlhaus.

Hyden, G. (1980) *Beyond Ujamaa in Tanzania: Underdevelopment and an Uncaptured Peasantry*, Heinemann, Nairobi.

Isard, W. (1956a) *Location and Space Economy*, Wiley, New York, NY.

Isard, W. (1956b) Regional science, the concept of the region and the regional structure, *Papers and Proceedings of the Regional Science Association*, Vol. 2, no. 1, pp. 13–29.

Isard, W. (1960) *Methods of Regional Analysis: An Introduction to Regional Science*, Wiley, New York, NY.

Isard, W. (1975) *An Introduction to Regional Science*, Prentice Hall, Englewood Cliffs, NJ.

Krebs, G. (1982) Regional inequalities during the process of national economic development: a critical approach, *Geoforum*, Vol. 13, no. 2, pp. 71–82.

Luttrell, W. L. (1971) *Villagization, Co-operative Production and Rural Cadres: Strategies and Tactics in Tanzanian Socialist Rural Development* (paper 71.11), Economic Research Bureau, University of Dar es Salaam.

Markusen, A. (1983) Regions and regionalism, in F. Moulaert and P. W. Salinas (eds.), op. cit., pp. 33–55.

Massey, D. (1984) *Spatial Divisions of Labour: Social Structures and the Geography of Production*, Macmillan, London.

Mathur, O. P. (ed.) (1982) *Small Cities and National Development*, United Nations Centre for Regional Development, Nagoya.

Max-Neef, M. *et al.* (1989) Human Scale Development; an option for the future, *Development Dialogue*, 1989: 1, pp. 5–81.

Moudoud, E. (1988) Regional development and the transition's dilemmas in the third world: the new debate and search for alternatives, *Africa Development*, Vol. 13, no. 3, pp. 57–75.

Moulaert, F. (1983) The theories and methods of regional science and regional political economy compared, in F. Moulaert and P. W. Salinas (eds.), op. cit., pp. 15–19.

Moulaert, F. and Salinas, P. W. (eds.) (1983) *Regional Development and the New International Division of Labour*, Kluwer-Nijhoff, Boston, Mass.

Myrdal, G. (1957) *Economic Theory and Underdeveloped Regions*, Duckworth, London.

Onimode, B. (1988) *A Political Economy of the African Crisis*, Zed Books, London.

Perroux, F. (1950) The domination effect and modern economic theory, in K. W. Rothschild (ed.) *Power in Economics*, Penguin, Harmondsworth.

Redclift, M. (1987) *Sustainable Development: Exploring the Contradictions*, Methuen, London.

Renaud, B. (1981) *National Urbanization Policy in Developing Countries*, Oxford University Press (for the World Bank).

Richardson, H. W. (1972) Optimality in city size, systems of cities, and urban policy: a sceptic's view, *Urban Studies*, Vol. 9, no. 1, pp. 29–48.

Richardson, H. W. (1973a) *Regional Growth Theory*, Macmillan, London.

Richardson, H. W. (1973b) *The Economics of Urban Size*, Saxon House, Farnborough.

Richardson, H. W. (1976) Growth pole spillovers: the dynamics of backwash and spread, *Regional Studies*, Vol. 10, no. 1, pp. 1–9.

Richardson, H. W. (1981) National urban development strategies in developing countries, *Urban Studies*, Vol. 18, pp. 267–83.

Richardson, H. W. (1987a) Spatial strategies, the settlement pattern, and shelter and service policies, in L. Rodwin (ed.) *Shelter, Settlement and Development*, Allen & Unwin, London, pp. 207–35.

Richardson, H. W. (1987b) Whither national urban policy in developing countries?, *Urban Studies*, Vol. 24, no. 3, pp. 227–44.

Riddell, R. (1985) *Regional Development Policy: The Struggle for Rural Progress in Low-Income Countries*, Gower, Aldershot.

Rodney, W. (1972) *How Europe Underdeveloped Africa*, Tanzania Publishing House, Dar es Salaam.

Rondinelli, D. A. (1983a) Dynamics of growth of secondary cities in developing countries, *Geographical Review*, Vol. 73, no. 1, pp. 42–57.

Rondinelli, D. A. (1983b) *Secondary Cities in Developing Countries*, Sage, Beverly Hills, Calif.

Rondinelli, D. A. (1983c) Towns and small cities in developing countries, *Geographical Review*, Vol. 73, no. 4, pp. 379–95.

Rondinelli, D. A., Dennis, A., McCullough, J. S. and Johnson, R. W. (1989) Analysing decentralization policies in developing countries: a political economy framework, *Development and Change*, Vol. 20, no. 1, pp. 57–87.

Rondinelli, D. A. and Ruddle, K. (1978) *Urbanization and Rural Development: A Spatial Policy for Equitable Growth*, Praeger, New York, NY.

Rostow, W. W. (1960) *The Stages of Economic Growth: A Non-Communist Manifesto*, Cambridge University Press.

Sack, R. D. (1974) The spatial separatist theme in geography, *Economic Geography*, Vol. 50, no. 1, pp. 1–9.

Sack, R. D. (1980) *Conceptions of Space in Social Thought: A Geographical Perspective*, Macmillan, London.

Sack, R. D. (1981) Territorial bases of power, in A. D. Burnett and P. J. Taylor (eds.) *Political Studies from Spatial Perspectives*, Wiley, Chichester, pp. 53–71.

Salinas, P. W. (1983) Mode of production and spatial organization in Peru, in F. Moulaert and P. W. Salinas (eds.), op. cit., pp. 79–95.

Save the Children Fund and Overseas Development Institute (1988) *Prospects for Africa*, Hodder & Stoughton, Sevenoaks.

Schiffer, J. (1981) The changing post-war pattern of development: the accumulated wisdom of Samir Amin, *World Development*, Vol. 9, no. 6, pp. 515–37.

Simon, D. (1989) Sustainable development: theoretical construct or attainable goal?, *Environmental Conservation*, Vol. 16, no. 1, pp. 41–8.

Slater, D. (1975) Underdevelopment and spatial inequality, *Progress in Planning*, Vol. 4, no. 2, pp. 99–167.

Slater, D. (1978) Towards a political economy of urbanization in peripheral capitalist societies: problems of theory and method with illustrations from Latin America, *International Journal of Urban and Regional Research*, Vol. 2, no. 1, pp. 26–52.

Slater, D. (1982) State and territory in post-revolutionary Cuba: some critical reflections on the development of spatial policy, *International Journal of Urban and Regional Research*, Vol. 6, no. 1, pp. 1–33.

Slater, D. (1987) Socialism, democracy and the territorial imperative: a comparison of the Cuban and Nicaraguan experiences, in D. Forbes and N. Thrift (eds.) *The Socialist Third World: Urban Development and Territorial Planning*, Blackwell, Oxford, pp. 282–302.

Slater, D. (1988) Capitalism and the regional problematic: why and how are peripheral societies different? (paper given at the IGU Working Group on Urbanization Conference, University of Melbourne, 15-16 August, published as Peripheral capitalism and the regional problematic, in R. Peet and N. Thrift (eds.) (1989) *New Models in Geography*, Vol. 2, Unwin Hyman, London, pp. 267–94.

Slater, D. (1989) Territorial power and the peripheral state: the issue of decentralization, *Development and Change*, Vol. 20, no. 3, pp. 501–31.

Smith, S. (1980) The ideas of Samir Amin: theory or tautology?, *Journal of Development Studies*, Vol. 17, no. 1, pp. 5–21.

Stöhr, W. B. (1981) Development from below: the bottom up and periphery inward development paradigm, in W. B. Stöhr and D. R. F. Taylor (eds.), op. cit., pp. 39–71.

Stöhr, W. B. and Taylor, D. R. F. (eds.) (1981) *Development from Above or Below? The Dialectics of Regional Planning in Developing Countries*, Wiley, Chichester.

Stöhr, W. B. and Tödtling, F. (1978) Spatial equity: some antitheses to current regional development strategy, *Papers of the Regional Science Association*, Vol. 38, no. 1, pp. 33–53.

Streeten, P. (1980) From growth to basic needs, in World Bank (ed.) *Poverty and Basic Needs*, Washington, DC.

Streeten, P. and Burki, S. (1978) Basic needs: some issues, *World Development*, Vol. 6, no. 3, pp. 411–21.

Sunkel, O. (1973) Transnational capitalism and national disintegration in Latin America, *Social and Economic Studies*, Vol. 22, pp. 132–76.

Susman, P. (1987) Spatial inequality in Cuba, *International Journal of Urban and Regional Research*, Vol. 11, no. 2, pp. 218–42.

Ternent, J. A. (1976) Urban concentration and dispersal: urban policies in Latin America, in A. Gilbert (ed.), op. cit., pp. 169–95.

von Freyhold, M. (1979) *Ujamaa Villages in Tanzania: Analysis of a Social Experiment*, Heinemann, Nairobi.

Weaver, C. (1981) Development theory and the regional question: a critique of spatial planning and its detractors, in W. B. Stöhr and D. R. F. Taylor (eds.), op. cit., pp. 73–105.

Wisner, B. (1988) *Power and Need in Africa: Basic Human Needs and Development Policies*, Earthscan, London.

Wong, S. T. and Saigol, K. M. (1984) Comparison of the economic impacts of six growth centres on their surrounding rural areas in Asia, *Environment and Planning A*, Vol. 16, no. 1, pp. 81–94.

World Bank (1981) *Accelerated Development in Sub-Saharan Africa: An Agenda for Action*, Washington, DC.

2

SPATIAL POLICIES AND UNEVEN DEVELOPMENT IN THE 'MARXIST-LENINIST' STATES OF THE THIRD WORLD

James D. Sidaway and David Simon

Introduction

> There is no shortage of available definitions of what socialism is, but there is a shortage of agreed definitions... . What is possible, of course, is to define the characteristics of a particular kind of socialism, existing or aspirational, preferred or disparaged, but this is a different enterprise (though often passed off as the same).
>
> (Wright, 1986, p. 19)

Regional policy and spatial planning have become universally accepted tools in the kitbag of development planners since the early 1970s. Manipulation of space in order to promote development had, of course, occurred previously. However, this was frequently an inadvertent or implicit outcome of aspatial, generally sectoral, planning rather than explicit policy.

Regional planning is today the subject of a prodigious theoretical and empirical literature, covering an array of paradigms and country experiences. One conspicuous feature, however, is the relative paucity of material on the large and diverse group of so-called socialist and Marxist-Leninist states in the Third World. Notwithstanding seminal work by David Slater (1975, 1982), John Friedmann (1980) and others, the edited volume by Forbes and Thrift (1987) is probably the first book-length treatment of the subject, at least in English. This lacuna is all the more remarkable in view of the explicitly state-centred approach to development implied by Marxism-Leninism and many variants of socialism. Concern with spatial inequality, territorial divisions of labour and regional relations of production should, therefore, be central to socialist development strategies.

Useful work has been produced on urban and regional issues in the 'developed' socialisms of the Soviet Union and Eastern Europe. Works such as French and Hamilton (1979), Musil (1980) and Demko (1984) contain a wealth of empirical information, but are not as theoretically informed as those that examine urban and regional themes in capitalist societies from the perspective of political economy. In the article by Demko and Fuchs (1979), and in Peet's (1980) reply, are the germs of what ought to be an interesting debate on how and why the social use of space in 'socialist' societies differs from the capitalist ones. But despite these promising openings, such debate has hardly begun. In a work claiming to be the 'first theoretical account, in the tradition of the new urban sociology' of the structural generation of urban inequalities under Hungarian socialism, Szelenyi (1983) provided a seminal study. In other East European states (Poland and Czechoslovakia) he was able to gain at best 'only unsystematic and often indirect evidence' (*ibid.* p. 7). Even more so than in Eastern Europe, the profound neglect of the Marxist-Leninist Third World probably lies in large measure with the sheer difficulty of undertaking critical research on such sensitive issues in countries that have either been engaged in countering concerted threats to their very existence or that have not yet imbibed the spirit of *glasnost*.

This chapter arises out of preliminary work on a study of spatial policy and uneven development in Africa's Marxist-Leninist states. It considers definitions and concepts before outlining the extent of our knowledge of these issues in 'socialist' and Marxist-Leninist states in the Third World as a whole.

Definitions and characteristics of the 'Marxist-Leninist' Third World

The 'Marxist-Leninist' label has today lost much of its original meaning in view of the number and variety of states claiming 'socialist' or even Marxist-Leninist ideologies. Moreover, many Third World countries possess features that have come to be associated with Marxism-Leninism, such as state ownership, authoritarian military or one-party structures and comprehensive planning, but make no claims to be pursuing a socialist transition. For example, the state and planning play an important role in most of the newly industrializing countries, which are overtly capitalist. In Africa, both statism (as the World Bank points out in its critiques), in the form of nationalized ownership and state control of agricultural prices and the weakness of the state are, to a degree, shared across the ideological spectrum. There are also seeming anomalies, such as the fact that the one-party state in the Ivory Coast (one of the most anti-communist countries in Africa) still today declares itself to be based on democratic centralism, inherited from the old alliance between the Rassemblement Democratique Africain (RDA)[1] and the French Communist Party. Furthermore, most states that claim to be Marxist-Leninist do not have the political features of 'direct' democracy or the level of 'development' that Marx and many Marxists today (particularly in the West) see as central to socialism.

So are there actually meaningful distinctions between Marxist and non-Marxist polities in the Third World? The answer is complex. Definitions such as that of White (1983, p. 1), which is used to introduce Forbes and Thrift's (1987) study of urban development and territorial planning in the socialist Third World are problematic. He argues (p. 3) that there are several structural features that differentiate a 'socialist' system from a capitalist one. First, the regime has 'broken ... the autonomous power of private capital over politics, production and distribution, abrogated the dominance of the law of value in its capitalist form, and embarked upon a development path which does not rely on the dynamic of private ownership and entrepreneurship'. A second requirement is that some fundamental transformations have been brought about to society and economy – 'most notably, the nationalization of industry, socialization of agriculture, abolition or limitation of markets, and the establishment of a comprehensive planning structure and a politico-ideological system bent on the transition to an ultimate communist society' (*ibid.*).

This approach (one used in many descriptions of transitional or socialist societies) is a rather static contrast of market and planned and/or state owned economies. Perhaps this is unsurprising given the highly *étatised* nature of many socialisms. But such a juxtaposition of planning versus the market in fact tells us very little about the class nature of different polities. For socialism, in Marx's vision, also meant the direct control by producers over political and economic decision-making. This does not exist in a meaningful sense in any of the societies that claim to be socialist, though it remains in many normative definitions of what socialism ought to be. In practice, socialism has come to mean many things, and several critical approaches to analysis of those states that claim to be in transition to socialism have evolved (see Holmes, 1986, pp. 379–401). At the very least, as Halliday and Molyneux (1981, p. 269) point out, the two components of any analysis of these societies ought to be 'on the one hand, a critique of their claim to have realized socialism; on the other, a concern with the mechanisms that govern them, with how these societies actually work'.

Though the Soviet Union has provided an important reference point for Marxist-Leninist states in the Third World, the latter are by no means all replications of the Soviet model. The states in the Third World that claim to be Marxist-Leninist are heterogeneous. Not only have the Chinese and Soviet paths diverged, but polities as varied as Cuba and Nicaragua, Mozambique, Kampuchea, Ethiopia and South Yemen have undergone social revolutions[2] and are ruled by parties who see themselves as Marxist.

Hence it must be stressed that individual Marxist-Leninist states are just that, and should not be reduced to a universal pre-ordained pattern. In particular, the future trajectory of many Third World socialisms is very uncertain. Some, for example, Grenada and Afghanistan, have been 'rolled back', collapsing under the combined results of external pressures and internal contradictions. As noted by Forbes and Thrift (1986, p. 24), for many of the recently founded socialist developing countries, 'the establishment of the state as determinant is often still problematic'. In addition, their economies are often externally dependent and still dominated by capitalist relations. The weak state is not hegemonic domestically and faces a hostile external climate. Recent dramatic events in Eastern Europe do make the future of Marxism-Leninism as a state ideology much less certain there. But what many observers see as the start of Eastern Europe's transition from Marxist-Leninist socialism to social democracy should not be projected onto the process of change in the Third World.[3] The dynamics of reform are not equivalent in Eastern European and Third World state socialisms. Indeed, to see in the contemporary changes in Marxist regimes an uneven but unilinear progression from 'socialist' totalitarianism to capitalist democracy is to fall into a similar trap to those who once claimed that all revolutions are made in Moscow and would, perforce, lead to governments that are Leninist in form and content. As Close (1989, p. 186) reminds us, 'During the nineteenth century a dazzling array of states – some democratic, others quite the opposite – were constructed around liberal principles'. We should expect a similar diversity of state forms around the ideologies of socialism – including Marxist socialism.

Kruijer (1987) has suggested the existence of four typical phases in the transition to socialism. The first two phases are associated with revolution and its consolidation. The third, state socialist phase, is characterized by centralized state control of production, powerful official bodies and the effective political subordination of the peasantry and workers. This corresponds to the Leninist model in Slater's (1987, p. 28) schematic division of 'the diverse faces of socialism' into 'reformist social democracy', 'radical nationalism as Third World socialism' and the Leninist model. Slater contrasts this with another approach, emanating largely from Gramsci, which emphasizes the absence of any pre-given automatic socialist consciousness rooted in class determination. According to Slater (*ibid.* p. 28),

> In a post-revolutionary situation such an approach to socialism would necessarily imply the guaranteeing of plurality and the continuing struggle for an effective abolition of all forms of alienation. In this sense the seizure of State power would not be taken as constitutive of an end of politics but rather one step, albeit an essential one, in the struggle to develop and securely root new forms of popular control and organisation.

Within this, he stresses that the spatial organization of the state and the wider social uses of space are domains for struggle.

Similarly for Kruijer (1987), with the advent of his fourth, democratic socialist phase, state power is significantly reduced and decision-making vested increasingly in the people. Kruijer acknowledges that the transition from state to democratic socialism may be exceedingly difficult, and will be vehemently opposed by the classes of officials and bureaucrats who control the authoritarian state machinery and thus stand to lose most. Ultimately, a second – and even more difficult – liberation struggle may be required to effect the transition. We will return to the theme of alternative visions of socialism and their related spatial forms later.

Spatial policies under 'Marxism-Leninism'

Whatever the exact status of Third World Marxism-Leninisms, there is relatively little literature on spatial planning in these states, and what exists has concentrated on the two, long-established examples of China and Cuba. As Bogdan Szajkowski

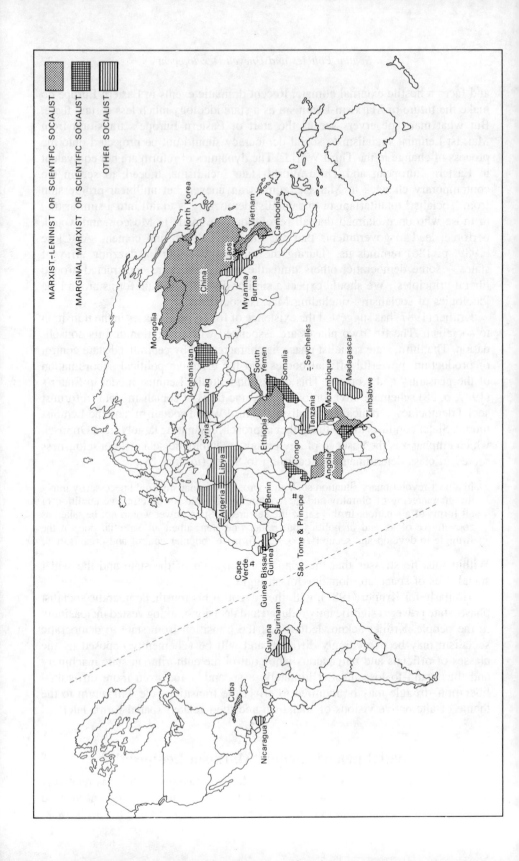

MARXIST-LENINIST OR SCIENTIFIC SOCIALIST

MARGINAL MARXIST OR SCIENTIFIC SOCIALIST

OTHER SOCIALIST

North Korea
Vietnam
Cambodia
Laos
China
Myanmar/Burma
Mongolia
Afghanistan
Syria
Iraq
South Yemen
Somalia
Seychelles
Madagascar
Mozambique
Zimbabwe
Ethiopia
Tanzania
Congo
Angola
Libya
Algeria
Benin
São Tome & Principe
Cape Verde
Guinea Bissau
Guinea
Guyana
Surinam
Cuba
Nicaragua

(1986, p. 5) notes in his editor's introduction,

> The study of Marxist regimes has not grown in proportion to the number of states claiming to be Marxist. For whilst for three decades after the Second World War, it was possible to distinguish eight regimes in Eastern Europe and four in Asia [China, Mongolia, North Korea, North Vietnam] which claimed adherence to the tenets of Marxism, currently at least 25 regimes claim to be Marxist.

In part it is because many of these 25, several of them hardly constituting states at all,[4] qualify only very marginally as states pursuing an effective 'transition to socialism', even by White's criteria, that observers such as Hyde-Price (1988) can argue that they have more in common with peripheral capitalist states (Figure 2. 1). This may be so for those where there has been no social revolution. However, the experience of those Third World states where social revolution has led to, or been led by, a polity that claims to be Marxist-Leninist diverges both from 'advanced socialism' and 'peripheral capitalism', despite their external dependency and weakness.

This weakness, reinforced by external pressures, also dictates that strategic considerations are paramount in the 'spatial policies' of many such states. That the activity of the state in space needs to be conceived in terms broader than the economic is evident when one considers, for example, Afghan or Angolan 'spatial policy'. Such extreme cases highlight the fact that even in less dramatic situations spatial policies generally have both explicit and implicit political functions.

As the work of Gore (1984) and Slater (1988) reminds us, regional planning does not occur in a political and economic vacuum. Yet there is still a disturbing tendency to accept uncritically the statements of planning agencies and governments and a corresponding failure to ask, in the words of D. M. Smith (1977, p. 7), 'who gets what, where and how' out of the process, or to conceptualize it rigorously within understanding of the state. It is necessary to consider all of the elements of the state in space. Even if attention is confined to only one such facet, namely regional planning, analysis ought to be linked to an understanding of all, semi-autonomous, realms of society and state activity – of economics, politics and civil society. In what follows some of the comparative literature on urban and regional planning in the 'Marxist-Leninist' Third World is reviewed with these themes in mind.

Forbes and Thrift (1987, p. 5) aptly note the paucity of literature on socialist developing country experience in the urban and regional arenas. In fact, none of the notable examples they list – Lo and Salih (1978), Honjo (1981), Renaud, (1981), Stöhr and Taylor (1981), Gilbert and Gugler (1982) – offers a comprehensive examination of the issue. The same applies to the case studies of individual country experience they cite. These cover Cuba and China, 'as well as some of the longer established socialist countries [sic] like Vietnam, Algeria, Tanzania and Guyana'.[5] Moreover only two of the case studies within the Forbes and Thrift volume itself, namely that by Mingione, and Slater's comparative essay on the relationship between alternative visions of socialism in Cuba and Nicaragua and the forms of spatial policy adopted, go beyond the descriptive presentation of empirical material.

Searching for shared patterns, Forbes and Thrift (*ibid.* p. 6) themselves claim that despite their heterogeneity, socialist developing countries do have common characteristics: a slower rate of urban growth accompanied by a reduction in the degree of urban primacy. They note that explanations of such 'polarization reversal' fall into two categories:

1. As an intended outcome that draws its strength from ideological convictions.
2. As unintended consequences of the pursuit by the state of more fundamental interests.

In the first case, the explanation rests on the assumption that socialism should generate within a society a radical alternative social organization and with it a radically different socio-territorial system. However they also note (*ibid.* pp. 9–10) that the foundations of a conception of spatial change in socialist construction are 'only very superficially articulated in the writings of Marx and Engels' and that Mao Zedong's anti-urbanism has had an enduring impact 'on a number of socialist developing countries, particularly those which, like China, had experienced a peasant revolution'. Within China itself, however, Forbes and Thrift suggest that restrictions on city growth have also been a means of minimizing investments in social overhead capital. Yet even this is not distinctive, since Gore (1984, p. 251) makes the same point of policies to reduce primacy and urban growth in capitalist states.

Forbes and Thrift (1987) concentrate mainly on the political functions of counter or restricted urbanization. They see the most sophisticated presentation of the political argument (for why 'socialism' leads to a slower rate of urbanization) in the model advanced by Murray and Szelenyi (1984). Its key proposition is that the main characteristic of the city is a large middle class or petty bourgeoisie. In socialist states the emergent class of bureaucrats and revolutionaries opposes the urban bourgeoisie and so the ensuing class struggle has anti-urban consequences, even though it is not primarily an anti-urban strategy. Forbes and Thrift (1986; 1987, p. 13) argue that the Murray–Szelenyi model is overly deterministic and/or reductionist and 'neglects critical processes in the realm of politics and civil society'. They stress the need, based on their study of Vietnam, to take into account such factors as warfare, external security threats and tensions over ethnicity. In particular, the impact on spatial strategy of the internal and external threats to security is stressed.

Forbes and Thrift (1986, p. 13) note, for example, that many 'socialist' developing countries 'have instigated population movements to threatened borders with consequent effects on the pattern of urbanization'. In all, they feel that, 'the list of effects of armed struggle [on patterns of urbanization] is long and constitutes a major research frontier' (*ibid.*).

Despite its neglect of such themes, Murray and Szelenyi's paper is indeed a useful starting point for analysis of urbanism in the transition to socialism. They distinguish between different patterns of urbanization that characterize the nature of urbanization and regional structures in different socialist countries and in different stages of their development. Four patterns are outlined:

1. The deurbanization stage.
2. The zero urban growth strategy.
3. Underurbanization.
4. Socialist intensive urbanization.

They argue that the stages represent an evolutionary path societies in transition to socialism have to follow in order to reach a 'mature' stage of socialism. It is acknowledged that not all societies have to pass through each stage, and they allow that the sequence might represent an 'ideal type'. Murray and Szelenyi describe the different patterns of urbanization in the transition to socialism in terms of associated changes in the occupational structure, rate and form of economic growth, and class structure and conflicts.

In summary, stage one, a relatively short initial stage, has been more pronounced in the Third World than in 'developed' socialist states, and is linked in Murray and Szelenyi's model with the social purpose of destruction of the urban petty bourgeoisie. Stages two and three are different patterns of urbanization representing two alternative routes of socialist economic development. The difference is based on conscious ideological choices – the Stalinist economic strategy leads to 'underurbanization', the Maoist strategy to 'zero urban growth'.

Of the ultimate easing of anti-urbanism in stage four, they note (1984, p. 102) two possible conclusions: 'that anti-urbanism is a necessary, but passing stage in the transition to socialism ... in order to reach a developed socialist society one has to dissolve the features of capitalist urbanism and as the socialist economic system begins to function gradually a new, socialist type of urbanism will evolve'. Alternatively,

> one could suggest that the easing of anti-urban fervour of socialist practices is an indication of the strengthening of restorationist forces. ... 'socialist intensive urbanization' evolves in the historic moment when major concessions are made by central planning to market forces and to private initiative, when a 'new petty bourgeoisie' is being created.
>
> (*Ibid.*)

Mingione (1987) also attempts to answer the question of why there is a tendency for slow and dispersed urbanization or even de-urbanization in socialist developing countries. He relates it to two of the goals pursued by socialist countries, namely the aim of increasing the surplus extracted from agriculture and using it to promote industrial development, and the commitment to maintaining a minimum level of survival for every inhabitant.

Slater (1987) goes beyond the economic realm to the political, investigating how different variants of socialist ideology are manifest in spatial organization. In comparing and contrasting the experiences of Cuba and Nicaragua, he is concerned not just with what has traditionally fallen under the domain of regional planning but also with the total organization of the state in space. Such an approach, which considers the functions of regional planning within the wider range of state activity in space, is rich: for potentially it enables us to see how different forms of spatial structure are related to different forms of polity.

What emerges is that just as the experience of socialism is heterodox, there is no single socialist pattern for the organization of space. What common patterns have been traced, namely the lower rate of urbanization and reduction in urban primacy, have had to be qualified with the observation that the patterns will be the result of a multitude of 'national paths to socialism' as they articulate with the global polity, economy and culture.

Theory and praxis: the way forward

Two broad themes are considered here: the diversity of spatial forms found in 'socialist' states, and what the examination of society and space under socialism contributes to our conception of uneven development in general.

Reflecting on the recent theoretical literature on urban and regional development, Slater (1988, p. 1) notes that the 'advanced capitalist' has been retained at the thematic centre of inquiry. He questions (*ibid.* pp. 14–15) the applicability of concepts like the 'hypermobility of capital', to peripheral capitalism:

> since although it corresponds to new trends within the advanced capitalist societies, and is also in evidence with relation to the expansion of transnational capital, it is much less relevant within most peripheral capitalist societies where capital is not as 'spatially indifferent' as elsewhere. ... [Hence] A research agenda formulated in the United States, Britain or the Netherlands, and quite naturally carrying with it a certain set of conceptual and methodological priorities ought not to be parachuted into a society like Tanzania or Colombia.

In parallel fashion, Watts (1989, p. 32) criticizes the Eurocentrism of the literature concerned with the restructuring of capitalism:

> It is a measure of the hegemony (and the parochialism) of the industrial, urban and regional restructuring literature that amidst all the heavy breathing over locality, reconstructed regional geography and so on, virtually no attention has been paid to the huge body of scholarship pertaining to the intersection of ethnography and political economy in the study of third world popular classes.

To this we should add that when the Third World has been taken into account, attention has centred largely on the peripheral capitalist states. But the socialist Third World is interesting not only for its own sake, and because it contains a significant proportion of the world's population, but also because the very primacy of the ideological and political facets to organization of space in many Marxist-Leninist states draws our attention to, and forces us to take account of, these realms.

The attempt by the post-revolutionary state to assert its hegemony and entrench itself nationally as well as to transform social relations will often encounter regionally based resistance – as in the Atlantic coast and Honduran border region of Nicaragua, and in Ethiopia, Angola and Afghanistan. The changing territorial organization of the state, particularly in the context of social revolution is, in itself, important.

As Slater (1988, p. 18) suggests, 'Not only does the establishment of regional development agencies need to be taken into account but also, and more systemically significant, the development and deepening of the territorial hierarchy of state power'. In this context the construction of new forms of political organization and mobilization in the post-revolutionary·states of the Third World, such as *poder popular* in Cuba and Mozambique and *Sandinismo* in Nicaragua, produce, when constituted in space, patterns different from those that preceded them. With a few exceptions such as Slater's (1987) and Egerö's (1987) respective comparisons of Nicaraguan and Mozambican spatial organization with that of Cuba, this is a wide and neglected theme.

The key point is that social revolution must be a sociospatial process, given that it develops in space–time. Be it from above or below, urban[6] or rural, social revolution unfolds spatially as well as sequentially. And the question of how the post-revolutionary state organizes itself in space reveals much about the difference that a revolution makes for society. For example, we need to examine the extent to which the post-revolutionary state has a high level of popular participation and autonomy in decision-making and the development of a social use of space that serves mass rather than just élite interests.

The Stalinist or quasi-Stalinist variant of Marxist-Leninist democratic centralism has exercised a tremendous influence over Marxist-Leninist praxis. The important point about the Leninist concept of democratic centralism is that, as Holmes (1986, p. 121, original emphasis) notes, 'the *noun* is centralism, the *adjective* democratic. In other words, the basic tenet is centralism, not democracy.' But it is also important to stress again that other 'socialist' conceptions of 'spatial problematic' do exist. Hence, Slater's (1987, p. 284) investigation using the Cuban and Nicaraguan examples of how the notions of popular hegemony, pluralism, devolution of power and regional autonomy differentiate Nicaragua from the more 'Leninist' Cuba.

Alternative visions of how society should use space are also part of the complex set of themes in other social revolutions. Thus Clapham (1988, p. 5) describes Northern Ethiopia where the Eritrean and Tigrean movements, who themselves are Marxist and have transformed social relations in the areas they control, are fighting for autonomy against the Marxist-Leninist state of Ethiopia:

> Northern Ethiopia is indeed a battleground between rival revolutions based on opposed (but equally revolutionary) principles and techniques; a centralizing and nationalist revolution originating in the towns, which has spread out to organize the countryside, and a decentralizing and regionalist revolution organized in the countryside seeking to surround and capture the towns.

Finally, it is important to record that within the pre-Stalinist Russia, alternative visions of how the new society should use space were articulated. Soja (1989, p. 89), drawing on the work of Kopp (1971), points to the existence of an avant-garde movement of city planners, geographers, and architects who worked towards achieving a 'new socialist spatial organization' to correspond with other revolutionary movements in Soviet society in the period from 1917 to 1925: 'Spatial transformation was not assumed to be an automatic byproduct of revolutionary social change. It too involved struggle and the formation of a collective consciousness. Without such effort, the prerevolutionary organization of space would continue to reproduce social inequality and exploitational structures.'

The innovative activities of this group were crushed and their memory erased by
Stalinism. Hence, for Soja (1989, p. 89), 'Productivism and military strategy came
to dominate spatial policy in the Soviet Union, all but burying the significance
of a more profound spatial problematic in socialist transformation'.

This leads us to our second broad theme, namely, what difference does the
operation of a socialist polity make to the mechanisms that produce uneven
development? The concept of uneven development has become a point of departure
for an increasing number of geographers. These analyses have come to conceive
of 'uneven development' not simply as a term of description, as a synonym for
regional disparity, but rather as a process – variously, a product of, or necessary
component of, the functioning of a capitalist economy. Such analyses claim to
show how uneven spatial structures have developed, and have then reproduced
themselves as an example of a sociospatial dialectic. Though there are important
differences, all of the work claims to illustrate how uneven development at any
scale is related to the processes inherent in capitalism.

Theorization on uneven development has advanced rapidly in the 1980s, from
a rather crude economistic determinism, more or less reading off patterns of
uneven development and even political responses to these, from the inner logic
of capital (Carney, 1980), through Harvey (1982), Massey (1984) and Smith (1984,
1986) to a more mature consideration of the links between agents and structures
in shaping patterns of uneven development, such as Duncan and Goodwin (1988),
Hudson (1988), Morgan and Sayer (1988) and Soja (1989).

But such description and theorization have largely ignored the experience of
'actually existing socialism'. Yet if, as has been argued here, many Marxist polities
are different from capitalist polities, even though not coinciding with Marx's
original vision of socialist society, then the neglect of human geography to include
an understanding of the production of spatial unevenness under socialism is a
significant omission.

The point has been recognized by David Smith (1987, p. 98), in a review of
works of Massey and Neil Smith:

'Like Massey, Smith confines his attention to capitalism. His justification is that to do
otherwise – to assert the generality of uneven development, is to reduce it to a universal
metaphysics; as such, "its critical, epistemological cutting edge is not only blunted,
but is potentially turned back on its user as a reactionary ideological weapon lurking
within the corpus of Marxism itself".'

But David Smith (*ibid.*) feels that

to understand the specific manifestations of uneven development under socialism, for
example, seems neither to evoke a universal concept nor to pose a danger to Marxism
as an analytical method (though some knowledge of the actual outcome of socialism
might have produced greater circumspection in a few passages which came close to
polemics). Sometime soon, the experience of socialism has to be brought into rigorous
relationship with the emerging theory of uneven development under capitalism.

In fact, in geography the question of how to deal with the 'alternative' system
of socialism introduced a complication that could most easily be resolved by
omission. Similarly, Corbridge (1988, p. 86) proposes that modern technologies

and forms of production themselves – in partial autonomy from their systems of organization and control – produce uneven development.

A similar point is made in de Gonzales's (1980, p. 2) work, where he suggests that some aspects of the relationship between society and space 'are articulated through technology – a factor which in many ways is fundamentally unchanged by a change in the mode of production'. But technology, of course, serves a purpose: the realization of surplus value. The key question then becomes, how is surplus realized in practice under socialism; how is it recycled; and how do such processes translate into spatial outcomes? Hence analysis of the structural mechanisms producing spatial differentiation or uneven development in polities that are not capitalist offers a potentially important contribution to our conception of the relationships between the 'spatial' and the 'social'. Radical geography, despite these and earlier observations, has hardly begun to consider these themes.

Environmental relationships also require incorporation into our conceptions. Just as space is now attaining its rightful place in social scientific inquiry so, too, do we need to understand the environment not merely as a set of passive resources and constraints but as a complex and dynamic system essential not only for development but ultimately for continued human existence. Socialist and Marxist-Leninist states have been at least as *dirigiste* in relation to the environment as their capitalist counterparts. This is explicable in terms of the strongly modernist, pro-industrialization ideology common in state socialist regimes, and the heritage of environmental irrelevantism that characterized much Leninist and neo-Marxist thought before the recent revival in environmental consciousness. Such views are also facilitated by the limitation of effective public criticism in many state socialisms. Happily, there are now encouraging signs that the situation is beginning to change as part of the global awakening to the environmental crisis. But a thorough reincorporation of the 'environmental' into social thought is an urgent priority.

Conclusions

Examination of the issues raised in this chapter reinforces the point that it is not central planning *per se* versus market *per se* that differentiates socialism from capitalism. Rather it is the class, gender, racial, political and environmental consequences of policy. We have maintained throughout that, however marginalized by Stalinism and its legacies, there are diverse visions of socialism, related to alternative conceptions of how society should use space and interact with the environment.

The past decade has seen a rich and vigorous debate on the difference that space makes to social processes. There has also been an increased recognition that behind seemingly abstract economic and technical debates on spatial planning policies and on environmental degradation, there are clashes of social interest. Such awareness and debate can and must be enriched by widening the parameters beyond 'advanced capitalism' into the periphery and to those polities that claim to be establishing alternative and more appropriate systems. In a recent work, Soja (1989, p. 243) insists on a human geography 'which recognizes that the organization of space is a social product filled with politics and ideology, contradiction and struggle, comparable to the making of history'. The manifestation in space of Marxist ideologies and 'socialist' politics is part of this variegated social product and demands our critical attention. The history of socialism shows that these patterns are going to vary, according to differing social, cultural, economic and political backgrounds.

Notes

1. The Rassemblement Democratique Africain (RDA) was an alliance of pro-independence parties in France's African territories. As an interterritorial grouping, it disintegrated in 1958.

2. There are sharp divisions as to when a transformation can be considered a 'revolution', for example, see Goldstone (1980). By social revolution we mean, after Barrington Moore (1967) and Theda Skocpol (1979), rapid, fundamental transformations of a society's state and class structures, accompanied and in part carried out by class-based revolts from below – though sometimes, for example, in Ethiopia, executed by a sector of the existing state apparatus.

3. The different forms of contradiction and hence different rationales for recent economic reforms in 'peripheral' and 'developed' socialisms are dissected by Littlejohn (1988).

4. They are not states in Weber's definition as that organization that can 'successfully claim the monopoly of the legitimate use of physical force in a given territory' (quoted in Miliband, 1973, p. 47).

5. Forbes and Thrift's (1987, p. 3) study includes case studies 'purposely drawn' from all categories of Third World 'socialism'. This is rather confusing since a fair number of the case studies are very marginally socialist, in terms of White's definition cited at the beginning of their book, e.g. Zimbabwe, Guyana, Nicaragua and Algeria.

6. Gugler (1988) argues that the urban has been the decisive arena in most Third World revolutions. For, however important rural conditions are, it is in the cities, as foci of state power, that the decisive confrontations have taken place. In fact, the urban character of contemporary revolutions can be overstated just as, more frequently, has been their rural character. Neither the 'urban' nor the 'rural' should be fetishized. The real point is the purposeful social action that makes a revolution unfold in space as well as time in the state undergoing revolutionary change.

References

Carney, J. (1980) Regions in crisis: accumulation, regional problems and crisis formation, in J. Carney, R. Hudson and J. Lewis (eds.) *Regions in Crisis*, Croom Helm, Beckenham, pp. 22–59.

Castells, M. (1972) *The Urban Question*, Edward Arnold, London.

Clapham, C. (1988) *Transformation and Continuity in Revolutionary Ethiopia*, Cambridge University Press.

Corbridge, S. (1988) Review of Neil Smith: *Uneven Development: Nature, Capital and the Production of Space*, Antipode, Vol. 19, no. 2, pp. 86–7.

de Gonzales, G. M. (1980) *Regional Planning under the Transition to Socialism*, Centre for Development Studies, Swansea.

Demko, G. (1984) (ed.) *Regional Development Problems and Policies in Eastern and Western Europe*, Croom Helm, Beckenham.

Demko, G. and Fuchs, R. J. (1979) Geographic inequality under socialism, *Annals of the Association of American Geographers*, Vol. 39, pp. 304–18.

Duncan, S. and Goodwin, M. (1988) *The Local State and Uneven Development*, Polity

Press, Cambridge.

Egerö, B. (1987) *Mozambique: A Dream Undone. The Political Economy of Democracy 1975–84*, Nordiska Afrikainstitutet, Uppsala.

Forbes, D. and Thrift, N. (eds.) (1986) *The Price of War, Urbanization in Vietnam, 1954–1985*, Allen & Unwin, London.

Forbes, D. and Thrift, N. (1987) Introduction, in D. Forbes and N. Thrift (eds.) *The Socialist Third World, Urban Development and Territorial Planning*, Blackwell, Oxford, pp. 1–26.

French, R. A. and Hamilton, F. E. I. (1979) *The Socialist City*, Wiley, New York, NY.

Friedmann, J. (1980) The territorial approach to rural development in Mozambique: six discussion papers, *International Journal of Urban and Regional Research*, Vol. 4, no. 1, pp. 97–115.

Gilbert, A. and Gugler, J. (1982) *Cities, Poverty and Development, Urbanization in the Third World*, Oxford University Press.

Goldstone, J. A. (1980) Theories of revolution: the third generation, *World Politics*, Vol. 32, pp. 425–53.

Gore, C. (1984) *Regions in Question: Space, Development and Regional Policy*, Methuen, London.

Gugler, J. (1988) The urban character of contemporary revolutions, in J. Gugler (ed.) *The Urbanization of the Third World*, Oxford University Press, pp. 399–412.

Halliday, F. and Molyneux, M. (1981) *The Ethiopian Revolution*, Verso, London.

Harvey, D. (1982) *The Limits to Capital*, Blackwell, Oxford.

Holmes, L. (1986) *Politics in the Communist World*, Clarendon Press, Oxford.

Honjo, M. (ed.) (1981) *Urbanization and Regional Development*, Maruzen Asia, Singapore.

Hudson, R. (1988) Uneven development in capitalist societies: changing spatial divisions of labour, forms of spatial organization of production and service development and their impacts on localities, *Transactions of the Institute of British Geographers*, Vol. 13, no. 4, pp. 484–96.

Hyde-Price, A. (1988) Review article: the international communist movement and third world Marxist regimes, *Manchester Papers on Development*, Vol. IV, no. 2, pp. 294–307.

Kopp, A. (1971) *Town and Revolution*, Brazillar, Paris.

Kruijer, G. J. (1987) *Development through Liberation: Third World Problems and Solutions*, Macmillan, London.

Littlejohn, G. (1988) Central planning and market relations in socialist societies, *The Journal of Development Studies*, Vol. 24, no. 4, pp. 75–101.

Lo, F. C. and Salih, K. (eds.) (1978) *Growth Pole Strategy and Regional Development Policy: Asian Experiences and Alternative Approaches*, Pergamon Press, Oxford.

Massey, D. (1984) *Spatial Divisions of Labour*, Macmillan, London.

Massey, D. (1987) *Nicaragua: Some Urban and Regional Problems in a Society in Transition*, Open University Press, Milton Keynes.

Miliband, R. (1973) *The State in Capitalist Society*, Quartet, London.

Mingione, E. (1981) *Social Conflict and the City*, Blackwell, Oxford.

Mingione, E. (1987) The urban question in socialist developing countries, in D. Forbes and N. Thrift (eds.), op. cit., pp. 27–50.

Moore, B. (1967) *Social Origins of Dictatorship and Democracy*, Penguin, Harmondsworth.

Morgan, K. and Sayer, A. (1988) *Microcircuits of Capital*, Polity Press, Cambridge.

Murray, P. and Szelenyi, I. (1984) The city in transition to socialism, *International Journal of Urban and Regional Research*, Vol. 8, no. 1, pp. 90–107.

Musil, J. (1980) *Urbanization in Socialist Countries*, Sharpe, White Plains, NY.

Pahl, R. (1977) Collective consumption and the state in capitalist and state socialist societies, in R. Scase (ed.) *Industrial Society: class, cleavage and control*, Allen & Unwin, London, pp. 153–71.

Peet, R. (1980) On geographic inequality under socialism, *Annals of the Association of American Geographers*, Vol. 70, pp. 280–6.

Renaud, B. (1981) *National Urbanization Policy in Developing Countries*, Oxford

University Press.

Skocpol, T. (1979) *States and Social Revolutions*, Cambridge University Press.

Slater, D. (1975) Underdevelopment and spatial inequality, *Progress in Planning*, Vol. 4, no. 2, pp. 99–167.

Slater, D. (1982) State and territory in postrevolutionary Cuba: some critical reflections on the development of spatial policy, *International Journal of Urban and Regional Research*, Vol. 6, no. 1, pp. 1–33.

Slater, D. (1987) Socialism, democracy and the territorial imperative: a comparison of the Cuban and Nicaraguan experiences, in D. Forbes and N. Thrift (eds.), op. cit., pp. 282–302.

Slater, D. (1988) Capitalism and the regional problematic – why and how are peripheral societies different? (paper given at the IGU Working Group on Urbanization Conference, Melbourne University, 15–16 August).

Smith, D. M. (1977) *Human Geography: A Welfare Approach*, Edward Arnold, London.

Smith, D. M. (1987) Review of Neil Smith's *Uneven Development: Nature, Capital and the Production of Space*, *Transactions of the Institute of British Geographers*, Vol. 11, no. 2, pp. 253–4.

Smith, N. (1984) *Uneven Development: Nature, Capital and the Production of Space*, Blackwell, Oxford.

Smith, N. (1986) On the necessity of uneven development, *International Journal of Urban and Regional Research*, Vol. 10, no. 1, pp. 87–104.

Soja, E. D. (1989) *Postmodern Geographies*, Verso, London.

Stöhr, W. B. and Taylor, D. R. F. (eds.) (1981) *Development From Above or Below? The Dialectics of Regional Planning in Developing Countries*, Wiley, Chichester.

Szajkowski, B. (1986) Editor's preface, in *Marxist Regimes*, Pinter, London, pp. v–vii.

Szelenyi, I. (1983) *Urban Inequalities under State Socialism*, Oxford University Press.

Watts, M. J. (1989) The agrarian question in Africa: debating the crisis, *Progress in Human Geography*, Vol. 13, no. 1, pp. 1–41.

White, G. (1983) Revolutionary socialist development in the Third World: an overview, in C. White, G. White and R. Murray (eds.) *Revolutionary Socialist Development in the Third World*, Wheatsheaf Books, Brighton, pp. 1–34.

Wright, A. (1986) *Socialisms*, Oxford University Press.

3

NATIONAL INTEGRATION:
CONTRASTING IDEOLOGIES IN GUINEA
Peter M. Slowe

Introduction

The integration of any state is hard to achieve when its population is ethnically diverse. When the transactions and affinities that define ethnicity ignore state boundaries, the state is left to create its own national equivalent of ethnicity. In other words, the state itself is left to generate its own transactions and affinities within its territorial boundaries – to create its own nationhood – and thus to become a nation state.

This is what happened in the West African state of Guinea (see Figure 3.1). Guinea had been an administrative area of about 100,000 square miles in French West Africa until independence in 1958 under the leadership of the trade union leader, Sékou Touré, and his Parti Démocratique de Guinée. The Republic of Guinea found itself with a population made up of about six million, 30 per cent each of Fulani, Malinké and Soussou, and a prospect of internal conflict. Development objectives were needed that would achieve a measure of internal integration as well as more conventional economic goals.

The long phase of development under Sékou Touré from 1958 to 1984 had very different objectives and methods, giving priority to integration, from those of the subsequent military government of Lansana Conté, giving priority to economic growth. Both development phases have important spatial implications, including different policies on ethnic regional development and state administration. The case of Guinea provides a good opportunity to examine the impact on development and change of two contrasting approaches to development applied successively to one African state.

Figure 3.1 Guinea: military regions, main urban centres and transport axes

Sékou Touré's Guinea (1958–84)

Introduction

The first leader of independent Guinea, Sékou Touré, and his party, the Parti Démocratique de Guinée, took the view that continued participation in international capitalism would make the integration of the Guinean nation state impossible. Touré's policies gave priority not to economic growth but to political, ethnic and class integration. This was a conscious and open priority. Touré wanted to avoid excessive dependence on world mineral prices and to avoid the overdevelopment of the capital city and its surrounding area that would inevitably have given a dominant role to the mainly Soussou French-speaking élite.

This is not to suggest that, in the late 1950s, Touré understood the full implications of dependency. Rather, the hostility of the French government when Guinea announced that it did not intend to participate in the French African Community forced Touré's hand and left him no alternative but to try to develop

Guinea in relative isolation to achieve his goal of integration. French capital fled, along with the business people and bureaucrats. So Touré no longer needed to argue that the dominance of French capital, with its concentration on production for export, would hinder integration. France cut off all aid and withdrew all its experts, which meant that change simply could not be piecemeal – a structural transformation of Guinea's economy and society would take place anyway, it was just a matter of whether Touré could take charge of the process.

Touré's idea was common to many developing states seeking some alternative to world capitalism, such as Nkrumah's Ghana and Nyerere's Tanzania. African agriculturalists, it was argued, once freed from the oppressive plantation or cash-cropping system, would produce more food for themselves and local sale and their incomes would increase. This would increase their demand for domestically produced manufactured goods, thereby stimulating new local industry. State intervention would foster new industrial enterprises, and the new industrial working class would benefit from increased employment; and the policy of maximizing consumer demand by redistributing income would ensure that they would benefit in real terms from economic growth. This would be the kind of sharp rupture, entailing deliberate state intervention in the organization of agriculture and industry, needed to turn a vicious cycle of decline and disintegration into a virtuous cycle of political and social integration.

A number of problems arose from this attempt at industrialization through import substitution. First, the scale of operation of the industrial processes, such as food-processing and textiles, required a bigger market than a single small state like Guinea could provide, resulting in expensive and underused capacity. Second, the initial demand profile, with its emphasis on the luxury tastes of a small élite, made more state involvement necessary than Touré had originally envisaged, because assumptions had to be made about future demand that depended on the success of state policies on distribution. Private investors were generally unwilling to take the risk, and so the state had to provide the initial economic stimulus. A strong central state was also needed to supervise the purchase of new technology, using revenue from mineral or agricultural exports, and the deployment of such technology in line with integration policies. Above all, a strong and interventionist central state was needed to carry out a policy that gave precedence to political, ethnic and class integration over economic growth in a free market.

Political integration

Touré aimed to achieve political integration through cultural change. On independence, Guinea took on all the paraphernalia of an independent nation state: a flag, an anthem and laws referring to state sovereignty. This was followed by a series of plots to destroy the Touré regime, starting with the 'Army Dissidents Plot' of 1960, organized by the French and culminating in a Portuguese-led invasion in 1970. While all this helped to establish Guinean identity within the new state boundaries, planned cultural change played the most important part.

Touré's view was that the rediscovery, development and dissemination of national culture were of the greatest importance in the process of political integration. He declared (1979, pp. 442–3) that

> The African personality cannot serve to mask anything which harms the interests of the African people, including purely economic interests, but it is the cultural development of African man which is the prerequisite for every other kind of development. His moral values, his intellectual capacity, his cultural characteristics are the prime manifestation and affirmation of his personality.

Political integration in the African state requires cultural change to the closest possible reflection of precolonial African values, and in the long run this is the only way to achieve suitable economic as well as political development in Africa. African culture 'is the active presence and the one driving force, the conscious instrument of a destiny on the path of progress' (*ibid.* p. 448).

The instrument for cultural change was a system of information flow that would educate Guineans to be independent of European ideas and educate the government in the ways and thinking of the ordinary Guinean in the village. Touré was faced with an overwhelming shortage of teachers and equipment, so education was reduced to its basic component, the transfer of information, data and ideas, from person to person, from place to place. As Mabogunje (1989) pointed out, good information flows, allowing for the free movement of data and ideas, are essential for development and change. He identified three stages of information flow: first, there has to be the physical infrastructure, next a model for the receipt of information and, only then, can the diffusion of the data and ideas themselves take place.

Touré used his limited economic resources to bolster the physical communications infrastructure (24.6 per cent of the Guinean budget in the first ten years after independence) and bought powerful transmission technology for his 'Voice of the Revolution' radio station; in this way, the physical basis for both aspects of information flow was prepared. Then he set out to prepare Guineans for an educational drive – in other words, to provide the model for the receipt of information. The key to his thought is revealing: 'The politics of teaching must precede the technique of teaching. It must be understood that the revolution prefers bad teaching, politically committed, devoted and honest, to perfect teaching which is reactionary and anti-popular' (Touré, 1961, quoted in Ameillon, 1964, p. 187).

There would be a two-way information flow. He would show Guineans how to express themselves and his cadres would listen and interpret, and convey ideas and data to the centre. The centre in turn would then provide revolutionary ideas and programmes that accorded with the priorities and preferences of the masses. It was not of the first importance either that the quality of the exchange was variable or that the quality of the data and ideas was sometimes banal. In this early period, it was the two-way, integrating model that mattered.

The flow of information from the villages to the centre reflected African cultural values and not the European objective of economic growth. Water was the first priority; education and health were far ahead of industry, commerce or agricultural development. On the basis of regular analyses of information on attitudes and requests for assistance from the new elected village councils, the Comités de Village, and local cadres of the Parti Démocratique de Guinée, Touré argued that the majority of Guineans had limited economic expectations – at least in the short term, perhaps just food and water, clothing and shelter in a subsistence environment. It was a practical first aim to ensure that these real priorities were met and not to be concerned about more elaborate ways of boosting economic growth. The budgets at the end of the 1964–70 Development Plan reflected African priorities by projecting over 30 per cent of government expenditure on information flow under the headings of 'Telecommunications, Training and Social Environment', compared with less than 10 per cent at the beginning of the Plan period (République de Guinée, 1965, 1971).

The devotion of substantial resources to direct communications between ordinary Guineans and their government, and especially between rural villages and the capital, brought about three different results. First, villagers came to identify in part with the state, alongside tribe and village. Second, villagers were encouraged to make decisions for themselves in their own physical and cultural environment, often leading to the articulation of alternative economic and social priorities to those proposed by the state's central planners. The argument that a greater awareness of one's own culture will alleviate a psychological dependence on external values and then instil a greater sense of purpose and motivation in development is now widely accepted (Davidson, 1982; Fyle, 1986), but in the late 1950s it was a very new idea. Third, the methods used ensured an effective information flow from villages to the government, so the government knew about local aspirations and could ensure its popularity more easily. Political integration was achieved and, at the same time, a great boost was given to the chances of keeping Touré and the Parti Démocratique de Guinée in power.

Ethnic integration

Ethnic or tribal groups in Guinea are concentrated in specific geographical areas. Policies of ethnic integration therefore entail spatial economic policies aimed at distributing the wealth of the state fairly between different areas. Indeed, ethnic equality was given such high priority by Sékou Touré that national economic development was sometimes sacrificed intentionally in favour of ethnic regional equality. For example, bauxite mining for export was slowed in 1959 to 60 per cent of its previous levels of output until iron ore extraction could be developed in the Kissi Forest region. Another example would be the pricing policy for rice, which favoured inland Malinké production, to make it possible to sell in the coastal capital, Conakry. The *Plan Septennal 1964–71* (République de Guinée, 1965) was in part a conscious exercise in ethnic integration, reducing anticipated economic growth to 0.8 per cent per annum and anticipating regional subplans audaciously biased against the capital city and export orientation and in favour of spatial equality. Although few of these subplans were even written, let alone implemented, the overall effect was to diminish the attractiveness of Conakry by creating relatively more wealth and opportunities in the countryside and provincial cities.

Under French rule the most important ethnic inequality (after the one between Europeans and Africans) was between the Soussou and other tribes. The Soussou, who came from the area around the capital, were the only tribe to be incorporated by the French, forming almost the whole of the emancipated *évolué* class. It was clear that to come to grips with this ethnic problem, a way had to be found to avoid the drift to the cities prevalent throughout the rest of the Third World. If real income could be equalized between the capital and the villages, the ascendancy of the Soussou would diminish.

The efficient two-way contact between the central government and local villagers, with no intervening regional tier of government, went a long way towards preventing some of the agrarian practices now recognized as mistakes in the rest of West Africa. For example, there was no interference with traditional land uses, crop rotation or fertilizer composting by central government agencies, as occurred, for example, in Sierra Leone, Ghana and Senegal. Guinea's growing rural population did not suffer from the deforestation and monocropping common right across Africa. The Fulani and Malinké concern for the preservation of forests and the lack of any widespread desire to shift to new modes of agricultural production emerged as village opinion and suited Touré well. The economic and political costs to the government of all except agreed minor improvements, such as well-sinking and the provision of basic education facilities, were avoided and few cultivators left the land for the city. The villages became increasingly attractive and the rural–urban shift was less than in other African states. Conakry grew by only 50 per cent between 1960 and 1980, whereas Freetown (Sierra Leone) grew by 1,000 per cent and Dakar (Senegal) grew by 450 per cent.

The oil-price crisis of 1973 onwards was also important in solving the problems of urban growth and Soussou ascendancy. Touré had continued to employ Soussou throughout the government; he needed them and they were loyal, but their privileges as government employees inevitably caused ethnic resentment. In 1973, real wages inevitably soon fell and, since it was in urban areas that regular wages were earned, these were the hardest hit. As in many West African capitals, despite the continuance of heavy subsidies on maize meal and rice supplied to the capital, junior officials were increasingly unable to support themselves and their families without also working in the informal economy or in agriculture. Fulani peasant smallholders, with a mix between cash and subsistence crops in Fouta Djallon, had higher disposable incomes by the mid-seventies than most Soussou in the capital.

Jamal and Weeks (1988, pp. 288–9) have shown convincingly that the disappearance of the rural–urban income gap is only a minor factor affecting rural–urban migration; the main attraction is for young men seeking some marginal increase in family income. Nevertheless, its disappearance gave Touré the chance to create centres of attraction in other parts of Guinea, which would be attractive to economic migrants of all sorts. Mining opportunities were created in the regions, with some of the largest developments in West Africa in Nzérékoré, Tougué, Boffa, Kindia and Mount Nimba. These labour intensive mines paid about double the wage of a middle-ranking government official and they attracted migrating labour away from the provincial capitals and Conakry. The population of Conakry stabilized from 1975 to the end of the Touré era in 1984. By 1980, the mean annual incomes of Fulani, Malinké and Soussou were about equal at the equivalent of approximately US $330 per head.

Class integration

The close links between class formation and the struggle for political domination in African states has been examined specifically by Murray (1967), Ake (1981) and Kafsir (1987). Murray says that new dominant classes are actually formed by accession to state power; Ake says that the economic and political bases must coincide if a class is to be dominant; and Kafsir argues that it is state patrimonialism that finally secures a dominant class in most African states. All three are clear that there is a crucial link between political and economic dominance within the state, normally associated with rule by a single party.

In Guinea, the objective of overcoming the class distinctions of the colonial and precolonial periods led to the early abolition of the chiefs and to the exclusion of business people, industrialists and the families of chiefs from any positions of rank in the ruling party. Touré's Parti Démocratique de Guinée became completely dominant within a year of independence (Adamolekun, 1976, pp. 77-87). No human activity fell outside the scope of the party, from baptism to food supply, from education to funerals. Everything was done 'for Party and Nation'. Nearly everyone in Guinea was a member of the party and had to attend meetings compulsorily with party cadres.

The dominant class swiftly became the party-cadre class in control of government and the parastatals, finance and a good deal of commerce and industry. By the mid-1960s, Sékou Touré had created a powerful élite with sole access to the cultural and economic machinery of the state (Yansane, 1984). He showed that he realized what was happening in *La Révolution Culturelle* (1969), where he wrote about the party-cadre class taking on the look of the bourgeoisie in colonial times and taking too little notice of the workers in the cities and the villagers in the countryside.

Touré's remedy was introduced in the 1973-8 Development Plan with its policy of devolution, reducing the state's tight control over commercial activities, collective farms and co-operatives. Above all, it devolved power to the villages and virtually removed the regional tier of administration. Villagers were given real power of decision over local agriculture, roads and schools, including forms of ownership, administration and economic activity. They could choose without reference to any other body, for instance, whether to use funds for road improvement or school building or for improving cash crops or subsistence crops, and they could run the local court. While the central government retained control of funding for major projects, weekly village Assemblées Générales, each with a part-time executive Comité de Village, ran the economic and judicial affairs of the village.

This devolution could take place in the villages without threatening Touré's government because the Parti Démocratique de Guinée retained overall control. Indeed, village organization was far less likely to threaten the party's position than regional organization. The party controlled the radio and other media; the party's patrimony extended throughout the economy and polity; the party's youth movement, the Jeunesse de la Révolution, and central party cadres on their workplace duties for a third of the year, were also a physical presence in the villages with access to the militia or army if they felt there might be any threat.

Devolution was safe for the ruling party class.

The village *assemblées* were, in fact, given local revenue-raising powers but they proved difficult to activate and hence ineffective. In practice, the villagers remained dependent on central government funding, which continued in most cases at about the same (very low) level as previously. About one third of revenue received from central government was earmarked unalterably for specific projects, about one third was unalterably for specific sectors, such as health or education but decisions could be made within the sectors, and the final third was for spending entirely at the discretion of the village. Central government could suggest uses, such as special subsidies for collective farm irrigation schemes or the salaries of teachers sent from Conakry, but these were not always followed. Although there were no clear performance criteria and almost no formal auditing, local party cadres acted as an informal but effective check on the *assemblées*. So, on the criterion of real local power over resources, Guinean devolution was only partial.

This is not to say, however, that Touré's government simply decentralized the state organization without giving any credence to local preferences – a common criticism of devolution policy elsewhere in the Third World (Harris, 1983; Slater, 1989; Simon, this volume). On the contrary, having established itself as the dominant class, the party élite no longer felt any need to enforce an expensive ideology on unwilling villagers. Between 1975 and 1980, about 80 per cent of collective farming practices were quietly dropped by the village *assemblées*, usually maintaining the original organizational shell but now run along more traditional co-operative or hierarchical lines. The same went for other local enterprises handed over to *assemblée* control. The local judicial system was still burdened with cases of anti-party ('anti-state') activity, but these were generally treated more leniently than before or referred to another court; in some parts of Fouta Djallon, attempts were made to interpret local law according to Islamic principles, and in other places local courts were used simply to pursue local feuds. Very little was done to give any clear direction to *assemblées*. Touré had destroyed the old classes and prevented the establishment of a new capitalist bourgeoisie. Now he wanted the new Guinean to emerge with new ideas under the protection of a new dominant class:

> It is here on the land, in the villages, that the fusion of energies takes place and here that the final sum of man's intellectual gifts is recorded. How can anyone think of excluding this source of thought, from this part of the human family next to the earth? In so doing, he may cut himself off from universal life?
>
> (Touré, 1979, p. 455)

To run the details of local administration from the capital had anyway been impractical, muddled and expensive. The village was a natural focus for devolution, the centre of organization of life in most of Guinea. Guinea's best-known writer, Camara Laye, in his most influential book, *L'Enfant Noir* (Laye, 1970), idealizes village life; while Mutiso (1974, p. 85) writes of African literature in general: 'those characters who are purely individualistic in African literature, no matter how much good they are doing, always end in tragedy ... the communal ethic is all pervasive'.

The village is therefore the logical level to which to devolve power, too small and perhaps too indoctrinated to pose a threat, yet the most meaningful geographically definable community for most Guineans. For administrative reasons, communities to which power was to be devolved had to have territorial definition, although the kinship ties that underlie many aspects of the village community tend to be geographically diffuse. The village is easy to define and the nature of devolution can be specified and measured.

The one-party system had declined by the late 1970s into a personality cult around Sékou Touré himself. It was the party élite surrounding Sékou Touré that discredited the system; kinship became nepotism; protecting the revolution became dangerous and antidemocratic paranoia. These were the failings of one leader and not a necessary failure of Touré's system.

The economic sacrifice for integration, however measured, was not great, compared with neighbouring states and other African states that attempted alternative development approaches (Table 3.1). For example, liberal economic policies have been relatively successful in their own terms in Côte d'Ivoire but not in Sierra Leone. On the other hand, Touré's Guinea and Nyerere's Tanzania provide models of integrated development with very slow economic growth. Extreme economic failure to the extent of Tanzania may inhibit integrated development. With completely inadequate goods and services, it has been impossible for Tanzania to secure the allegiance of the masses to the state, making Nyerere's radical approach to African nation building very difficult to achieve. The state that succeeds economically may be in a better position to survive political, ethnic and class divisions, since an abundance of goods and services would provide something substantial to share between the various rival groups. Touré's Guinea avoided extreme economic failure through the exploitation of substantial mineral resources and it also had a more closed economy and society for a longer period than any other African state, making comparison more difficult for the majority of its citizens. As Adamolekun, writing eighteen years into Touré's rule, commented (1976, p. 185):

Table 3.1 Mean percentage annual economic growth in Guinea, compared with other states, in the last five years of Touré's rule (1979–84)

State	Growth (% p.a.)	GDP/head, 1984 (US$)
Guinea	2.72	335
Côte d'Ivoire	0.84	679
Ghana	-1.14	411
Liberia	-1.88	396
Mali	0.38	135
Senegal	1.82	363
Sierra Leone	2.50	307
Tanzania	0.16	243
Zambia	0.84	427

(*Source:* OECD, 1989.)

Although some theorists have drawn attention to the important ways in which economic progress can in fact hinder the process of nation building, only a few writers have gone further to suggest that considerable progress can be achieved in the process of nation building despite a slow rate of economic development. The point that we wish to make here is that the Guinean experience is an example of a state that has achieved some progress in the process of nation building despite a slow rate of economic development.

Lansana Conté's Guinea (1984–   )

Introduction

Sékou Touré was buried on 30 March 1984. Three days later a senior army officer, Colonel Lansana Conté, led a military coup. He immediately announced the introduction of a free-market system giving absolute priority to economic growth within the international capitalist system. Within a few weeks, the Parti Démocratique de Guinée was disbanded along with the Jeunesse de la Révolution and its associated militia. Two months later, the village *assemblées* and *comités* were also disbanded.

It would be a mistake to look for any profound explanation for the sudden abandonment of the previous development goals by the new government. Touré's increasing autocracy in his later years and the unpopularity of his family members who squabbled to succeed him paved the way for a coup. Once in power, Conté had to destroy the Parti Démocratique de Guinée as a power base to secure his own position. In disbanding the party, he destroyed the political integration process, which depended on party cadres to function, and the planning system that had implemented the rest of Touré's policies. Conté also achieved some quick and simple popularity by abandoning Touré's cultural policy and opening Guinea's borders to Western culture.

Conté made it clear on becoming President that he was going to pursue liberal economic policies, to make economic growth his main goal and to invite foreign investment and assistance. It was an unplanned adoption of conventional economic growth objectives which the international community quickly rewarded with promises of economic aid, encouragement for private investment and, in the case of France, military aid as well. The chosen model was the well-known one for all economies failing to achieve growth: a draconian programme of reorganization under the aegis of the IMF, with rapid devaluation, a severe reduction in public spending, state withdrawal from the productive sector and an end to nearly all state subsidies.

The state banks were effectively taken over by private French banks within six months, and this was just the start of an almost complete withdrawal of the state from the productive economy. The Guinean economy was liberalized at a time when decentralization in the form of privatization was at the height of economic fashion. Goods and services were divided up into open-access, joint-use and private categories (Caire, 1988, pp. 20–4). Open-access goods and services, such as ground water, were consumed jointly and simultaneously by many people; it was hard to exclude anyone unable to pay, and privatization was not a sensible option. Joint-use or 'toll' goods and services, such as public transport, which were charged for, were possible but not essential to privatize. All other goods and services were privately purchased and the state would stop producing them; the free market was to operate and privatized production would be led by demand. There was no room in this analysis for the links between ability to pay, real possibilities to choose and the distribution of income. The free market was deemed to be apolitical and to act automatically in the interest of society as a whole, while political activity was to be confined to the government, which would sometimes need to use 'threats, pressures and punishments', rather than 'educational and persuasion techniques' to impose this kind of programme (Rondinelli, McCullough and Johnson, 1989, p. 80).

The approach of Conté's Comité Militaire de Redressement National to political integration was to rely on the inherent strength of Touré's nation state until democracy could eventually be restored. Their approach to ethnic integration was *ad hoc* and, in direct contrast to Touré's deliberate policy of state control of investment, no attempt was made to intervene in the locational decision-making of foreign investors. As for class integration, the state has come to be dominated by the military rather than either a party or a new economic bourgeoisie. The solidity of the inherited nation state and a certain amount of luck have enabled Conté's Guinea to enjoy the benefits of greater economic freedom without some of its dire potential costs.

Political integration

The emphasis Touré had laid on traditional African culture and his discouragement of the free market had led a number of villages to undertake very little trade. In the Fouta Djallon Hills and in the Malinké areas further inland, villagers tended to breed livestock only for local consumption and no longer grew cash crops, such as coffee and pineapples, concentrating instead on local food crops like cassava and maize. Away from the cities and main transport routes, villages became disengaged from the state – they had cut themselves off from it (Azarya and Chazan, 1987, pp. 126-9). When Conté abolished the party, most of the formal contact also disappeared. This form of disengagement has been largely eliminated during the first six years of the military regime and it is now calculated that as few as 3 per cent of villages in Guinea (down from about 50 per cent at the end of the Touré era) engage in no agricultural trade beyond their immediate area (Bureau Économique du CMRN, 1989).

Another form of disengagement from the state was the informal economy. Black markets, smuggling and corruption were parallel systems in operation in Guinea when Conté assumed control. This informal economy brought about by artificial pricing, trade restrictions, scarcity and an overvalued currency practically disappeared when rice and maize meal subsidies were removed early in 1985 and the currency was devalued.

The most blatant form of disengagement from the state under Touré was self-imposed exile. Conté's government has made strenuous efforts, mostly unsuccessful, to persuade exiles to return, especially since they often have useful skills and speak English or French. Here, as with the other types of disengagement, the Conté government has moved towards re-engagement whereby Guineans are encouraged to participate in one economy and live in one territory.

Conté's government has, however, discontinued the two-way flow of information between villages and the centre. Conté has simply kept the one-way flow of information by maintaining radio and television transmission at a high technical quality. African values are of no interest to a regime whose principal objective is economic growth regardless of village opinion. Neither the *assemblées* nor the party-cadre system have been replaced, so there is no official tier of administration at village level and even the 32 prefectures have few administrative functions. Military administration is highly centralized, with just five regional ministries and military provinces (see Figure 3.1). These are Conakry prefecture, Guinée Forestière (the remote southern forests, administered from Nzérékoré), Guinée Maritime (the Soussou coastal belt, administered from Kindia), Haute Guinée (the Malinké area up to the border with Mali, administered from Kankan) and Moyenne Guinée (Fulani Fouta Djallon, administered from Labé).

On 2 October 1989, the 21st anniversary of independence, Conté announced that a new democratic constitution would be drawn up for a referendum. Two political parties will be allowed and there will be a general election for a civilian government in 1994. Introducing a democratic constitution to an integrated nation state with popular participation could signal a new political era in Guinea. On the other hand, Conté's announcement included this ominous note: 'The man in uniform is a full citizen and the adoption of a constitution does not exclude him from the affairs of the state.' This has been widely interpreted as meaning that Conté will manipulate any elections, like Samuel Doe in neighbouring Liberia, to ensure that he continues in power. Conté is Soussou, from the traditionally dominant ethnic group, so the future of the Guinean nation state under those circumstances would depend very heavily on Conté's ability and willingness to maintain policies of ethnic integration.

Ethnic integration

Within the African state,

> the axis of measurement for development indicators is relative rather than absolute, between parts of a state or between African and European states. ... Development is about people; individuals and their institutions. It relates essentially to issues of the quality of life of the citizen, to the concept of the equality of persons, and to the equity or social justice of territorial organisation.
>
> (Fakolade, 1989, p. 360)

It is the measure of the relative development of Guinean ethnic regions populated mainly by Soussou, Fulani or Malinké that is the most likely to be compared by Guineans.

Touré's ethnic regional policies were immediately abandoned by Conté. In the first two years of the new government, the rate of capital investment in Guinea as a whole increased by 114 per cent. Of total new investment, 56 per cent was in Conakry and the surrounding area, 31 per cent was outside Conakry but in bauxite mining or ancillary activities and only 13 per cent (an actual reduction of about 25 per cent over the last two years of the Touré regime) in other activities in other places. Consequently, the population of Conakry grew from 300,000 to about 750,000 between 1984 and 1987. Agricultural investment fell back initially by some 60 per cent under the new regime, whereas investment in non-bauxite mining activities, particularly gold and diamonds, increased by some 80 per cent (Cheveau-Loquay, 1987; Bureau Économique du CMRN, 1988).

The economic liberalization that enabled these developments to take place was part of the programme agreed with the IMF. The reduction in the size of the state sector was an important part of the programme, not just from privatization but also from trimming the civil service and closing parastatals. Early in 1986 in Conakry, 10,000 civil servants and some 30,000 parastatal employees were dismissed by the Soussou-dominated military government. Malinké people were over-represented five times in the sackings with the obvious result that the new poor are nearly all Malinké or Fulani, and the sharp cuts in agricultural investment have left them little alternative but to exist on the fringes of urban life.

Whereas Touré's regime targeted ethnic and geographical integration, Conté's regime has aimed primarily at economic growth. Touré consciously sacrificed overall economic growth in favour of ethnic integration, but the Soussou have been the clear political and economic beneficiaries of Conté's regime. Conté has risked, in particular, the wrath of the Malinkés of Guinea's interior in favour of his own Soussou people based around the capital city.

A good deal of the overseas aid Guinea receives helps to compensate for the disintegrating side effects of liberalization, especially by concentrating on Fulani and Malinké rural areas. With 600 million f. in 1987 and 580 million f. in 1988, Guinea has been running second after Côte d'Ivoire as a recipient of French aid aimed especially at the development of new primary resources. Japan granted Guinea a loan of 5.5 billion Y on IDA terms in 1987 and 600 million Y in grants for rural development as a part of the special IMF Facility for Africa in 1988 and again in 1989. the IMF–World Bank programme is for $115 million per annum for three years, two thirds of it to finance development projects and one third to help the balance of payments. Under Lomé III, the EEC committed, in February 1988, a 70 million ECU programme centred on Malinké areas, and three similar loans were converted to grants following the 1988 Toronto economic summit.

In rural development, £280 million has been allocated for the period 1988–91 to develop Guinea's roads. An emergency programme prioritizes a huge primary-network rehabilitation scheme. The idea is to asphalt 1,200 km of road, extending the Conakry–Kankan highway to the remote Mamou–Faranah–Kissidougou–Nzérékoré stretch (Figure 3.1), and to build 1,900 km of rural tracks as well. The whole programme should mean a dense, properly constructed primary and secondary network by 1993, making for easy access to the small peasant farms in remoter production areas and removing a major barrier to Guinea's mineral and cash-crop exploitation. The production of rice, coffee, oil and palm, mangoes, pineapples and bananas is already expanding quickly on the large plantations of Guinée Maritime and more easily accessible areas of Moyenne Guinée. But it is for mineral exploitation that infrastructural aid has had its most significant effect on the economies of the Fulani and Malinké regions.

Conté's regime has gone out of its way to make itself attractive to international investors and, ironically, it may well be they who promote ethnic integration, albeit unintentionally. Large scale investment is nearly all associated with mining and ancillary activities in the remoter parts of the country.

Guinea has immense mineral wealth, including the purest bauxite in the world. Under the auspices of the multinational, Alusuisse, about 14 million tonnes a year of this ore are being extracted from three major sites, two in Fulani Moyenne Guinée and one in Soussou Guinée Maritime. At this rate of exploitation, known reserves should last for 500 years. There are 6.5 billion tonnes of iron ore reserves, including very high-grade (70 per cent plus) ore from Mount Nimba in the remote Guinée Forestière. Ore exports via Liberia are now planned to start at 6 million tonnes in 1990, the high quality making exploitation profitable even at current low world prices. Rural Guinea also produces diamonds, mined and marketed by the French multinational, Arédor. These are of such purity that demand from the jewellery industry is always heavy. The estimated reserves are 400 million carats, two thirds of them of gem quality. Finally, Guinean gold is historically recorded as having formed the basis of the region's wealth from the fifteenth to the eighteenth centuries. Gold is still mined, mostly by craft gold workers operating individually or in small co-operatives, but Union Minière, a French-based consortium, is now starting to organize the industry.

In summary, foreign investment in Guinea is inadvertently promoting ethnic equality and happens to counterbalance the Conté government's pro-Soussou bias. Aid has as its main focus rural development, especially in the remoter, and therefore non-Soussou, areas. Multinational investment is heavily concentrated on mining. To meet its economic objectives, Conté recognizes the need for such investment. Ethnic integration is a hardly noticed by-product, but it has resulted in relatively little ethnic tension under Conté, the exceptions being a series of Malinké riots in November and December 1987 (Andramirado, 1987) and a Fulani riot at Labé in October 1989 (Diallo, 1989).

Class integration

Military rule is not the same as authoritarian rule. On the contrary, regimes like that of Lansana Conté can be politically liberating from civilian one-party states. Additionally, the military, as a powerful state organization, is as capable of integrating the state as any single political party. There are, of course, political, ethnic and class rivalries in Conté's army, but its self-image is still that of a pivot guaranteeing state autonomy, not tolerating challenges from groups within the state (Odetola, 1982, p. 100). In Guinea, the military took the place of the party élite who had filled the same role. Under Touré, the Parti Démocratique de Guinée held autonomous state power and now the military does the same.

The Conté government's liberal economic policies, which have given free rein to market mechanisms, are most clearly expressed in the three-year Development Plan 1986–9 (Caire, 1988). The continuation of a public sector in industry and the provision of social services of any description are seen as likely to be inflationary and wasteful. Their existence under the old regime is blamed for the failure to create a Guinean business community to take over the privatized textile mills or port companies – there were few new entrepreneurs except in peripheral services and retailing, who had been there under Touré. A bourgeois class still does not exist at the end of the plan's three years and therefore many of the inequalities associated with an indigenous capitalist sector do not yet exist either. The civil servants and parastatal management of the Touré era have not so far applied their managerial experience to private enterprise as was hoped, and the educational and economic infrastructure for the early development of indigenous entrepreneurship in industry, as opposed to real estate and retail trade, is still lacking. If it does emerge, it could challenge military domination and create the kind of polarization between local élites and the rest of the population common elsewhere in Africa – but so far it has not and this sort of class disintegration has not yet occurred.

Conclusion

The contrast between the aims of the Sékou Touré and Lansana Conté regimes could not be greater. Touré's priority was the integration of an African state; Conté's has been maximum economic growth within the international capitalist economy. Sékou Touré's ideology gave indigenous African culture an outstanding opportunity to find expression in the modern world, and he provided an important model of integrated development appropriate to Africa. Conté's liberalization provides a model of economic and social change in an integrated nation state.

Touré regarded Guinea's mineral wealth as the basis for self-sufficiency, even isolation, politically and culturally as well as economically, and used it to put brakes on development in Soussou Guinée Maritime and to accelerate development in remoter parts of Guinea. Conté has opted simply for the maximum exploitation of resources to achieve the fastest possible growth in a privatized economy in agreement with the IMF. The signs are that the political integration Touré achieved through cultural development may have been sacrificed. Continued ethnic integration, however, should enable Guinea to survive as a nation state, to retain its national ethnic identity or nationhood, rather than just to survive as the legal entity of a state.

Guinea, since independence, has had over thirty years of political, economic administrative and spatial planning under two contrasting regimes. The implications for regional development have included the controlled development of regional mineral resources under Touré and unfettered exploitation under Conté, political and administrative decentralization with functional centralization under Touré but the reverse under Conté. Despite a dramatic change in the direction and style of its development in 1984, Guinea remains a poor but relatively equitable nation state.

On a continent where development efforts after independence failed because the misconceived strategy was to make a dash for modernization by copying rather than adapting Western models (International Bank for Reconstruction and Development, 1989), Touré's Guinea provides an object lesson. Yet it also had the faults that the World Bank recognizes in the same report (*ibid.* p. 18): poorly designed public investments in industry, too little attention to agriculture, too much intervention in areas where the state lacked expertise. It was Guinea's good fortune that, with the process of nation building nearly complete, it was able to adopt policies of economic liberalization. Some states, such as Nigeria and Kenya, have tried to pursue growth and integration simultaneously; others, like Tanzania and, to some extent, Ethiopia and the Somali Republic, have had policies giving priority to integration but have not had Guinea's resources relative to population. The uniqueness of Guinea's development lies in the sequence of events that enabled it to pursue, first, integration and then growth.

References

Adamolekun, 'L. (1976) *Sékou Touré's Guinea*, Methuen, London.

Ake, C. (1981) *A Political Economy of Africa*, Longman, Harlow.

Ameillon, B. (1964) *La Guinée, Bilan d'une Indépendance*, François Maspéro, Paris.

Andramirado, S. (1987) Conté face au défi intérieur, *Jeune Afrique*, 16 December, pp. 28–30.

Azarya, V. and Chazan, N. (1987) Disengagement from the state in Africa: reflections on the experience of Ghana and Guinea, *Comparative Studies of Society and History*, Vol. 29, no. 2, pp. 106–31.

Bureau Économique du CMRN (1988) *Revue Économique et Financière*, Bureau Économique du CMRN, Conakry.

Bureau Économique du CMRN (1989) *Revue Économique et Financière* (draft), Bureau Économique du CMRN, Conakry.

Caire, G. (1988) Guinée: deuxième gouvernement de la deuxième République, an deux, *Mondes en Développement*, Vol. 16, no. 62, pp. 15–33.

Cheveau-Loquay, A. (1987) La Guinée: va-t-elle continuer à négliger son agriculure?, *Politique Africaine*, Vol. 25, no. 1, pp. 120–6.

Davidson, B. (1982) Ideology and identity: an approach from history, in H. Alavi and T. Shanin (eds.) *Introduction to the Sociology of 'Developing Societies'*, Macmillan, London, pp. 435–56.

Diallo, S. (1989) Violentes émeutes a Labé, *Jeune Afrique*, 13 November, pp. 10–11.

Fakolade, A. (1989) Cosmopolitanism vs provincialism, in A. I. Asiwaju and P. O. Adeniyi (eds.) *Borderlands in Africa: A Multidisciplinary and Comparative Focus on Nigeria and West Africa*, University of Lagos Press, Lagos, pp. 353–60.

Fyle, C. M. (1986) African culture and higher education: the Sierra Leone situation, *Africana Research Bulletin*, Vol. 15, no. 2, pp. 3–19.

Harris, R. L. (1983) Centralization and decentralization in Latin America, in G. S. Cheema and D. A. Rondinelli (eds.) *Decentralization and Development: Policy Implementation in Developing Countries*, Sage, London, pp. 183–202.

International Bank for Reconstruction and Development (1989) *Sub-Saharan Africa: From Crisis to Sustainable Growth, a Long Term Perspective Study*, Washington, DC.

Jamal, V. and Weeks, J. (1988) The vanishing rural-urban gap in sub-Saharan Africa, *International Labour Review*, Vol. 127, no. 3, pp. 271–92.

Kafsir, N. (1987) Class, political domination and the African state, in Z. Ergas (ed.) *The African State in Transition*, Macmillan, London, pp. 45–60.

Laye, C. (1970) *The African Child*, Fontana, London.

Mabogunje, A. L. (1989) *The Development Process: A Spatial Perspective* (2nd edn.), Unwin Hyman, London.

Murray, R. (1967) Second thoughts on Ghana, *New Left Review*, Vol. 42, pp. 26–36.

Mutiso, G.-C. N. (1974) *Socio-Political Thought in African Literature*, Macmillan, London.

Odetola, T. O. (1982) *Military Regimes and Development: A Comparative Analysis of*

African States, Allen & Unwin, London.

OECD (1989) *Latest Information on National Accounts of Developing Countries*, OECD, Paris.

République de Guinée (1965) *Plan Septennal 1964-71*, Ministère d'État, Conakry.

République de Guinée (1971) *Revue du Développement Économique*, Ministère d'État, Conakry.

Rondinelli, D. A. (1981) Government decentralization in comparative perspective: theory and practice in developing countries, *International Review of Administrative Science*, Vol. 47, no. 2, pp. 133–45.

Rondinelli, D. A. (1982) The dilemma of development administration: uncertainty and complexity in control oriented bureaucracies, *World Politics*, Vol. 35, no. 1, pp. 43–72.

Rondinelli, D. A., McCullough, J. S. and Johnson, R. W. (1989) Analysing decentralization policies in developing countries: a political-economy framework, *Development and Change*, Vol. 20, no. 1, pp. 57–87.

Slater, D. (1989) Territorial power and the peripheral state: the issue of decentralization, *Development and Change*, Vol. 20, no. 4, pp. 501–31.

Touré, A. S. (1969) *La Révolution Culturelle*, République de Guinée, Conakry.

Touré, A. S. (1979) *Africa on the Move*, Panaf, London.

Yansane, A. Y. (1984) *Decolonization in West African States with French Colonial Legacy*, Schenkman, Cambridge.

4

SUBNATIONAL DEVELOPMENT
AND PLANNING IN
PACIFIC ISLAND COUNTRIES[1]

Hidehiko Sazanami and Roswitha Newels

Introduction

The 1970s and early 1980s witnessed a widespread interest in the development problems of small, remote, island and landlocked countries. This interest manifested itself in a series of conferences, publications, the creation of a great many expert groups and, last but not least, the adoption of a United Nations programme of action for least-developed, landlocked and small island economies. An almost surprising quantity of materials has been published on a broad range of macro-economic, sectoral and local-level community-development issues and problems faced by island nations in their attempts to develop, (Brookfield and Hart, 1971; Selwyn, 1975; Jalan, 1982; Dommen and Hein, 1985; United Nations, 1985; Cole and Parry, 1986; Srinivasan, 1986; Browne and Scott, 1989). Little systematic and comprehensive research and analysis is, however, available with regard to the spatial structure of island economies and the complexities of spatial development and planning in the context of island socioeconomic systems.

For example, it is surprising that the importance of rural–urban socioeconomic linkages and the role of the urban centres or small service centres in rural development have, with the exception of Ward and Ward (1980) been analysed and recognized primarily with regard to rural–urban migration issues. For useful reviews and analysis of regional development and planning issues and problems in small Pacific island nations, see Gunasekera (1982), Chandra and Gunasekera (1984), United Nations Centre for Regional Development (1982, 1984); Standingford (1985) and Overton (1987).

Development planning and management remain comparatively recent undertakings in most of the Pacific countries (Gunasekera, 1982). Planners are experimenting with a range of development models, approaches and planning methods. Whilst significant progress has been achieved during the past decade in national, sectoral and project planning, progress in subnational (regional/local) planning and management has been much slower. In response to political pressure and existing interregional and rural–urban socioeconomic disparities, the development plans of all the Pacific nations cite the need for spatially more balanced development and growth as a prime development objective.[2] However, no example of comprehensive subnational development planning and management exists in any of the island economies. National plans usually contain only implicit recognition of the spatial structure of the economy and, at best, provide for superficial treatment of development programmes at the subnational level. This chapter reviews selected aspects of the practice and problems of subnational development planning in Pacific island nations, concluding that there is a need to strengthen subnational development and planning within a multilevel development planning and management framework where planning for development is viewed as vision, and dialogue based on participation, co-ordination and integration.

Initial attempts at introducing a spatial component into the overall development and planning process can be traced back to the early 1970s. In most cases subnational development and planning efforts, whether termed decentralization, basic needs provision, integrated rural development, community and/or outer-island development, lacked integration within the framework of overall national, sectoral development planning and management. Planning directed towards subnational development consisted and still generally consists of a range of piecemeal approaches, projects and programmes that can do no more than alleviate existing disparities in levels of living and income between the rural and urban areas of nations in the short to medium term, if at all. It is also important that planning is not viewed as just a public administration exercise and/or in isolation from the dynamics underlying development in the island nations.

Many Pacific island nations are 'microstates' with populations below 100,000. Consequently, they face many development constraints, including the countries' overall smallness in size; varying degrees of internal fragmentation; small domestic-market size; dual, highly open and vulnerable economies; undeveloped or underdeveloped rural–urban linkages; high per capita costs in the provision of basic socioeconomic infrastructure; in many cases poverty of resources; and both internal and external communication problems.

Corbin (1985) summarizes the major socioeconomic characteristics of Pacific microstates as being size, openness, remoteness and fragmentation, level of development, dependence and dominance of the public sector. The following generic characteristics tend, therefore, to be closely related to planned policy objectives and strategies:

1. Major regional disparities in levels of development exist among the constituent islands but predominantly between the cash economies of the urban centres and the traditional economies of the outer islands.
2. The economies have a high degree of physical fragmentation with weak

transport and communication linkages, the improvement of which seldom justifies private sector investment, thus necessitating heavy subsidization.

3. A high proportion of GDP is generated in the traditional sector that often engages over 70 per cent of the economically active population, and small organized sectors exist in agriculture, fishing, services and manufacturing.

4. All of these economies have a limited fiscal base, low levels of public-sector saving, heavy dependence on import tariffs and often on such sources as philatelic sales for the generation of domestic revenue.

5. The governments are dependent on official foreign transfer for almost all development expenditure and often a significant part of recurrent expenditure.

6. The recurrent expenditure and capital account expenditure of the public sector constitute a major proportion of total GDP.

7. The public sector is the principal source of wage employment in the economy.

8. The economic base of the economies is narrow, with dependence on a single or limited range of primary commodities for export.

9. Export earnings are unstable and disproportionate to the costs of imports of necessary capital goods and current levels of imports of intermediate and consumer goods, resulting in an endemic disequilibrium in the external account.

10. Unrequited transfers in the form of remittances from relatives abroad are often a major source of cash income.

11. There is dependence on a donor-supported expatriate workforce to assist in offsetting a lack of technical and managerial skills.

A commonly held view is that a few judiciously selected, well-designed development projects, together with efficient public sector investment programming, rather than an ongoing planning operation, are a largely adequate and efficient method for the development and management of small island economies. (Allen and Hinchliffe, 1982). The question could be raised as to whether there is any requirement for systematic, comprehensive and integrated subnational planning and management. We argue that these island economies differ in scale but not in kind from other developing nations and thus face similar constraints, problems and needs, including the requirement for a spatial approach to planning and development. Subnational development and planning in the island nations also cannot be viewed in isolation from the need to progress towards necessary national political and economic integration. This development policy issue is of particular relevance to these newly independent states marked by cultural-linguistic diversity and strong local identities that are often reinforced by the geographical distance between constituent island groupings (regions) of countries.

Many of the Pacific island nations have only recently achieved independence (Papua New Guinea, 1975, Solomon Islands, 1978, Tuvalu, 1978, Republic of Kiribati, 1979, Republic of Vanuatu, 1980). The institutions that have then been set up, usually provided with personnel who have only limited planning experience, are faced with the task of planning and managing development. The Prime Minister of Tuvalu, Bikenibeu Paeniu, in a recent interview (1989) quite pointedly stressed that

> it will be the aim of my government to continue to promote a sense of national identity and help our people to realise that we are Tuvaluans, not just from Nukulaelae, Nanumea,

Vaitupu... . It is in the collective interest to have a solid base from which to work and that means we must continue to support developments on the capital.

In helping the islands we must be very careful not to disintegrate the nation. For example, in the case of the Kingdom of Tonga, the northern group of the Niuas (Niuatoputapu and Niuafo'ou), is some 530 kilometres away from the capital island of Tongatapu. In the case of the Republic of Kiribati, distances between the constituent island groupings are extreme. The Republic of Kiribati comprises some 31 islands and atolls. They are widely dispersed within three distinct 200 mile Exclusive Economic Zones (EEZs) that cover 3 million km^2 of ocean. Kiritimati Island in the east, for example, is about 3,870 km from Banaba Island in the west (see also, Bonnemaison and de Deckker, 1990).

In search of development and planning models

It is surprising that the considerable body of empirical research and, to a lesser extent, theoretical work on a vast range of development issues in small island nations has not yet generated, to our knowledge, either an island-nation specific spatial-development theory or spatial development and planning models. The transferability of conventional regional development and planning theory and approaches, such as growth pole theory, export base theory, gravity models and so on is questionable. The existing regional development and planning theories and concepts all incorporate useful elements but do not identify and isolate the factors that define the political, socioeconomic, geographical and cultural development dynamics of the island nations (Tabriztchi, 1988). The transferability and/or relevance of regional development and planning theory have been widely debated (e.g. Kuklinski, 1980, Richardson, 1980, Richardson and Townroe, 1986, Morrison, 1988, Tabriztchi, 1988).

Two development planning models have, however, recently been developed by the United Nations Department of Technical Co-operation for Development (DTCD). These are the Macroeconomic Simulator for Island Economies (MESIC) and a complementary model by DTCD's Development Planning Advisory Branch in close collaboration with the United Nations Centre for Regional Development, namely, the Regional Economic Model for Island Countries (REMIC).

The REMIC model incorporates the generic structural characteristics of the island economies and proceeds within the institutional framework and technical capabilities of planning agencies, including the national planning and sectoral agencies and subnational governments. The principal objectives pursued in modelling regional socioeconomic development in island economies were to develop a broad conceptual framework and to ensure its accessibility to a broad range of planners and policy-makers. The model finds some antecedents in growth pole models and export base models.

The most prominently researched and widely advocated policy, planning and management framework for inducing subnational development in the island countries is decentralization (Standish, 1979; Berry and Jackson, 1981; Conyers, 1981; Larmour and Qalo, 1985; Premdas, 1988). Decentralization is viewed as the panacea to the 'real' problems of development and it appears that policy-makers assume that decentralization measures will ensure increased equity in development among subnational units. However, the centralization versus decentralization issue has been greatly oversimplified and very inadequately analysed within the development context of island nations.

In innovative research, the process of centralization versus decentralization is modelled within the REMIC institutional framework. This framework addresses questions concerning the strategic allocation of planning responsibilities for regional planning and development between the national and subnational governments. The forces governing the centralization versus decentralization process are identified and the concepts of organizational strategy space and centralization versus decentralization decision space are introduced in an attempt to take account of institutional, administrative and political factors.

As in other parts of the world, decentralization is often viewed as a means to achieve objectives such as maximizing the national rate of growth, achieving more spatially balanced growth through the reduction of socioeconomic disparities among constituent regions of a country and inducing greater levels of participation in development and planning. When discussing decentralization in the context of small-island countries, apart from taking into account the generic structural characteristics of island economies, we must bear in mind that we are not dealing exclusively, or even primarily, with endogenous development processes. An overwhelming proportion of development resources at the national and subnational levels are provided by foreign aid and will continue to be derived from this source for the foreseeable future. In practice, decentralization is currently confined to the decentralization of administrative authority that can contribute to the achievement of an increased rate of growth at the national level where development resources are deployed on the basis of economic-efficiency criteria. This, it is believed, increases administrative efficiency and, therefore, the productivity of government bureaucracy; increases local participation in development planning and management over and above levels that would otherwise be available; and directs the allocation of resources so that they reflect more closely the development preferences of local populations and the national population, thus enhancing the overall rate of development programme implementation.

Subnational governments in the Pacific, however, lack the skilled staff and finance to deliver the wide range of services they are legally required to provide. There is an urgent need to strengthen subnational-level institutional and administrative capacity to foster subnational-level participation in development and planning efforts. At present, decentralization is too indiscriminate, often premature, *ad hoc* and not accompanied by planning and general management training. Not only does it provide no alternative to a comprehensive, ongoing planning operation but it also hampers the development process more than assisting it. Furthermore, decentralization as a policy device for promoting territorial equity has been shown to be neither logically justified nor empirically verified (Corbin and Newels, 1988).

Subnational planning observed: planning by default[3]

The institutions concerned with subnational development exist at both the national and subnational level. All of the institutions have only recently been established, have limited general management and planning experience and are currently strengthening their capacities.

National constitutions provide for various types of government at the subnational level. The Constitution of the Federated States of Micronesia, for example, enacts national, state and municipal levels of government and provides for revenue sharing between the national and state governments. In the Solomon Islands, the Constitution enacts provincial governance and it is the Principal Provincial Government Act 1981 and the Provincial Government Amendment Act 1986 that outline the functions and responsibilities of provincial governments as well as their relations with central government. Generally, constitutional provisions and/or decentralization Acts vary considerably in terms of their degree of specificity regarding the functions transferred to lower levels of government. In several cases, they lack clear delineation of national versus subnational government functions and responsibilities. At subnational level, such functions and responsibilities refer to general administration, and subnational governments are empowered to levy certain taxes such as head taxes, business licences and fees, and liquor, dog and similar licences.

The participation of subnational-level administrations in the planning and management of development remains constrained by the paucity of locally derived resources and the weak institutional and administrative capacity of these governments. The comprehensive and systematic preparation of local government area development plans has been undertaken in hardly any of the Pacific island countries. Where attempts at such preparation have been made, the approaches have varied considerably from country to country. The subnational demand for development resources is generally expressed through individual project submissions to national sectoral agencies for their funding. The Solomon Islands embarked upon an extensive provincial plan preparation exercise. These provincial plans, however, are not coterminous with the national plan, which will limit the scope for their systematic integration into and programming within the national sectoral plans (Corbin, 1987a). In Tuvalu, outer island development plans were prepared during 1985; while in the Kingdom of Tonga, regional plans were prepared through external assistance for two of the Kingdom's five regions. In the Republic of Vanuatu, the National Planning and Statistics Office undertook the country's first systematic regional development and planning survey during 1983. A series of meetings with local government councils was conducted, during which local development preferences and priorities were identified. During 1985–6, the Planning Office and the Department of Local Government organized yet another series of regional development workshops with the objectives of identifying project and programme priorities for the country's Second National Development Plan and of fostering local level participation in overall plan preparation. Some 6,000 local level projects were identified, some of which were incorporated into the national plan.

Of the national institutions with responsibilities associated with subnational development, the Ministry of Home Affairs or Interior and the National Planning Office are the key institutions. These planning offices, in most instances, are not equipped with specific regional planning capacity. Sectoral agencies of the national governments would normally provide substantive support to, and assist, local governments. However, the sectoral agencies remain in their build-up stage and have consequently gained only limited experience in planning and project implementation. Often sectoral agencies, such as health, education, agriculture and fisheries, maintain direct representation in local government regions through their extension officers (Tables 4.1 and 4.2).

Table 4.1   Total number of civil servants seconded to local government areas by sectoral agency, Republic of Vanuatu (June 1988)

Agency	Number of civil servants
Health Department	218
Public Works Department	52
Education Department	n. a.
Labour Department	7
Police, prisons, immigration	98
Agriculture Department	64
Rural water supply	27
Co-operative Department	50
Ports and Marine Department	25
Customs Department	10
Fisheries Department	14
Post and telecommunications	16
Civil Aviation Department	11
Rural Lands Department	4
Survey Department	7
Radio Department	1
Statistics Department	1
Meteorological Department	12
Total	617

Note

The staff seconded to local government by the Department of Local Government, Ministry of Home Affairs, are not included in this listing. It could be estimated that the ratio of central government seconded staff to local government employed staff, excluding education, is roughly 6:1.

(*Source:* Public Service Department, Republic of Vanuatu.)

Table 4.2   Education Department, Republic of Vanuatu: employees by region
            (excluding HQ staff)

Region	Total	Teachers	Civil servants	Other
1. Banks/Torres	50	45	—	5
2. Santo/Malo	190	161	5	24
3. Ambae/Maewo	111	91	4	16
4. Pentecost	116	104	—	12
5. Malekula	178	160	5	13
6. Ambrym	63	58	—	5
7. Paama	21	20	—	1
8. Epi	43	35	—	8
9. Tonga/Sheperds	44	40	—	4
10. Efate	361	286	4	71
11. Tafea	183	171	4	8
Total	1,360	1,171	22	167

(*Source:* Public Service Department, Republic of Vanuatu.)

Local government staffing thus consists primarily of staff seconded to local government areas by central government. The lack of planning and implementation capacity at local government level constrains subnational development. Currently, the local governments rely primarily on technical assistance provided to them by the central government's sectoral agencies. The lack of intersectoral communication and co-ordination is a further impediment to integrated and co-ordinated subnational development. Other agents active in local level development and planning are the national development banks, primary co-operative societies, community development organizations and church organizations. All perform important development functions and assist in the identification, planning, funding and implementation of community- and village-level projects and programmes. The main problem areas regarding the effective implementation of subnational initiatives thus relate to the lack of effective institutional structures and personnel to plan and implement development projects; limited project/programme co-ordination capacity at both national and local levels due to the absence of effective management-information systems; and limited institutional capacity of local governments to plan and implement development programmes and projects.

The main instrument for subnational development is the national plan, including sectoral projects and programmes. All of the plans incorporate a chapter on regional development and/or rural development, rural business development, community development, co-operative development and so forth. The sectoral project and programme summaries are only presented in terms of their functional and economic classification. Regional development chapters tend to be primarily descriptive and are limited in scope, listing some priority areas of activity by region. The sectoral agencies at the national level currently provide a major component of total development activities in all subnational government areas.

Several planning procedures, essentially in project planning and management, are elaborated. The existing planning procedures generally lack clear definition and are not always coherent. Planning procedures often comprise a plethora of elements and forms, at times even contradictory, and are all but conducive to the integration of national, sectoral and subnational planning.

Usually, the major portion of direct financial assistance received by subnational governments emanates from national government grants to local governments. Development resources are overwhelmingly provided for by foreign aid and this external dependence will continue. In all countries there is a range of financial instruments available for the funding of subnational-level programmes and projects. Comprehensive systems of national development budgeting, where established, are in their formative stages. The annual development budget is rarely used to implement subnational development policy, but merely serves the purpose of distributing project funds on an *ad hoc* basis. Since the development budget has to be politically approved on a year-to-year basis, subnational development consequently depends on the uncertainties of the political decision-making process. These uncertainties are heightened by the dependence of development budget funding on aid grants and loans. A high proportion of the budget funds is negotiated on an annual basis by the national aid management agency, which, in the process, has to deal with problems associated with synchronization and the dependence of subnational planning on donor preferences. There are no current criteria for the rational allocation of development resources between subnational units to ensure the systematic implementation of the governments' subnational development policies and strategies. Hence, development resources are allocated on a project-by-project basis. In the island states without a strong central planning agency, project programming depends on the negotiating power and lobbying of sectoral and subnational administrators. The co-ordination of sectoral project programming is constrained due to the often limited planning capacity of national sectoral agencies. Consequently, it becomes difficult even to set sectoral priorities. At the subnational level, regional programmes consist of a variety of small-scale projects and no subnational resource allocation criteria are established at either national sectoral or subnational level. The subnational or local-level development resource allocations consist primarily of inputs in kind (land, labour, materials) to development projects. No development budgets are prepared by subnational governments (Corbin, 1987a). Subnational development projects are allocated through project profiles to the central planning agency without referring to standard project design criteria, resulting in a high degree of misallocated, unco-ordinated projects. A low project-implementation rate is further limited by technical and personnel constraints and there is no systematic assessment of the maintenance and staffing cost impact of projects on the national recurrent budget. This is important to stress, as the current revenue position of local governments can best be described as weak, and it is the central government transfers that finance local level recurrent expenditure. In the Solomon Islands, for example, 77.3 per cent of all provincial government recurrent expenditure and 99.3 per cent of all provincial capital expenditure are met by central government. Central governments, however, are increasingly experiencing problems of maintaining current levels of intergovernmental transfers.

Subnational development and planning: beyond the local perspective

Being small is not always beautiful and does not exempt islands from experiencing spatial development problems and regional disparities in socioeconomic development. The need for a spatial approach to development in the island nations is primarily the result of their opening up to, and irreversible integration into, the world socioeconomic and political system. This brings about rising domestic expectations and needs. All the island countries are experiencing increased demand for the provision of social services and for modern goods that only a few decades ago could not even be found in the more traditional, closed societies. The islands have rapidly become part of the system of international relations and their futures are affected by the changes and developments in this system. These international developments, changes, movements of capital and labour almost immediately affect the course of development at all levels of the island economies.

Fairbairn and Tisdell (1984), for example, argue in favour of expanded production of subsistence, traditional goods as a means of sustained, more self-sufficient economic development in the island countries. However, such policy proposals are believed to be unrealistic as the process of integration of the island economies into the world system is irreversible.

The challenge for policy makers and planners is to co-ordinate rising expectations and needs with limitations in resource endowment, productive capacity and institutional capacity; cultural preservation; and conservation of fragile ecosystems. The main policy problem is how to manage this necessary and inevitable international integration, to induce or sustain national integrity and integration so as to allow for phased, self-sustaining socioeconomic development and growth by meeting popular demands and expectations at the same time. Addressing this problem requires comprehensive participatory planning and management that explicitly incorporates the spatial dimension of development and planning. Any micro-approach to subnational development and planning such as rural development or administrative decentralization alone ultimately bears the risk of leading to fragmented political and socioeconomic development, reducing further an already small domestic market size, increasing the economies' vulnerability to external shocks and reducing already limited scope for economic diversification. Only a macro-approach to subnational development and planning can stimulate socio-economic development whereby measures such as community development and rural development are seen as integral components of an overall, integrated, national spatial development and planning policy.

Integrating subnational development and planning in a multilevel planning and management framework

Planning for development of the island economies is still essentially sectoral at this stage. This approach involves a series of problems as sectoral plans do not deal systematically with the spatial aspects or impacts of their development activities. Sectors negotiate over development resource allocation on an exclusively intersectoral basis and there is no coherent analysis in terms of regional needs as opposed to the overall potential of the space economy. We argue that only a co-ordinated and integrated planning process based on participatory dialogue can be conducive to the phased implementation of long term development objectives such as self-sustained and spatially balanced development and growth. In the long run, decentralized planning and the planning-by-projects-only approach produce fragmented development that is not based on a vision for future development but the mere haphazard result of individual development initiatives. Subnational development and planning are proposed to be undertaken within the framework of multilevel development planning and management based on the principles of maximum participation, co-ordination and integration so as to attain long-term, self-sustaining socioeconomic development. Currently, planning is undertaken exclusively by the national administration with little or no planning activity at the subnational level. The operationalization of a fully fledged multilevel planning and management system is viewed as the ultimate objective in the phased and evolutionary development of the planning system.

The UNCRD definition of multilevel development planning and management goes beyond a mere public administration approach that simply analyses the division of functions between different tiers of government. Although suggesting primarily an operational sense and purpose, this approach seeks to analyse, co-ordinate and integrate development problems and the hierarchy of development need; development objectives; and political, social, economic and cultural reality based on an initial understanding of the dynamics of socioeconomic development, in a participatory framework, involving all levels and all agents of the economy. Despite sometimes wide geographical dispersion, the small population base of island nations should facilitate such a comprehensive approach to participation. Also, due attention ought to be given to existing traditional structures of communication, information dissemination, consensus building and participation in development activities.

Initially, the need is to strengthen national, sectoral and subnational government planning and general management capacity simultaneously. This ongoing, comprehensive strengthening at all levels of an existing institutional system should induce lasting subnational-level participation in the process. This upgrading requires training in areas such as general administrative management capacity, project planning and management, local level plan formulation and implementation and the like. Even today, the subnational-level institutions are usually provided with little if any systematic administrative, planning and management training. They are often not even provided with basic information concerning existing planning legislation or national-plan documents and are neither aware of, nor trained in, prevailing planning procedures. Training and information dissemination regarding established planning procedures, and the role and functions of central government's agencies, are thus essential components of a comprehensive subnational-level training package. Local level management training must have a comprehensive focus as it is necessary to create awareness of the rationale underlying planning and management, i.e. identifying and implementing the vision of people regarding the long term development path of their community, region and nation. Such local or subnational training needs to be conducted in conjunction with training for sectoral agencies at the national level so as to increase the efficiency of the line agencies in delivering projects, programmes and technical assistance to subnational levels. The role of the central planning office as a co-ordinating and rationalizing body in the overall national, sectoral and subnational development and planning process also requires strengthening.

Such comprehensive training constitutes the starting-point in the phased implementation of a multilevel development planning and management system and aims at fostering participation at all levels. Subnational participation in development and planning has two distinct dimensions – the need to induce greater levels of community, village-type development participation and the need to foster subnational-level government involvement in national, sectoral planning. Both require systematic integration into national sectoral planning and management. The integration process would initially involve a comprehensive analysis of the development situation and perceived development needs of subnational units in the form of simple, straightforward subnational development profiles. These profiles are the base component in the preparation of the sectoral development plans, projects and programmes and the identification of a national, macrospatial development and planning policy. The sectoral and spatial resource-allocation mechanisms and criteria will then be established on the basis of the adopted national policy. In this way, the national development plan with its sectoral projects and programmes will more adequately reflect subnational development potentials, constraints and needs. The objective of enhanced community and subnational government participation in overall national, sectoral planning and management will also be achieved.

Conclusion

The phased implementation of a multilevel development planning and management system based on participation, co-ordination and integration is viewed as a management tool assisting the gradual achievement of long term development objectives such as self-sustaining socioeconomic development. The prevailing approaches of planning by projects or financial programming alone are considered to be essentially short-to-medium-term development management tools that cannot adequately address long term development perspectives and goals. Multilevel development planning and management, in the context of small island nations, are furthermore greatly facilitated by the presence of relatively cohesive, communication-based and highly interactive social structures.

The presence of a complex set of development needs, together with considerable economic and planning constraints, necessitate an integrated, comprehensive approach to development and planning in island countries. The limited planning and management capacity of national planning and sectoral agencies as well as of subnational governments has been identified as being the main constraint to socioeconomic development. Limited planning and management capacity, however, cannot be an excuse for no planning. To quote the Chief Secretary of the Republic of the Marshall Islands, 'no plan is a plan for failure'. The absence of comprehensive, co-ordinated planning and management means the absence of a long term vision for the socioeconomic development path of a nation and leads to fragmented, short term management of development. Development planning and management at all scales are to be viewed and used as essential tools for development and not as some administrative exercise separate from the actual development process. The absence to date of realistic, broadly based conceptual frameworks for a macro-approach to spatial development and planning in island economies has led to a range of *ad hoc*, micro-approaches to subnational development that are not integrated into national sectoral planning. To a great extent, this can be explained by the lack of interdisciplinary research and understanding regarding the dynamics and the forces underlying development in the context of small island economies.

Hence, the role of multilevel development planning and management is not limited only to strengthening horizontal and vertical integration of institutional, community-based and private-sector planning and management activities but also aims equally at comprehensive, i.e. interdisciplinary, management of the consensus-based long term development path of a nation. The proposed approach to the planning and management of subnational development in small island countries within a multilevel framework whereby all planning activity is based on an understanding of the long term dynamics of socioeconomic development, might well be of wider relevance than just to small island developing countries.

Notes

1. The opinions expressed are solely those of the authors and do not necessarily reflect the positions of their respective organizations in any way.

2. See, among others, the Federated States of Micronesia, First National Development Plan (1985–9), 'Ensure an equitable distribution of the benefits resulting from development both among the States and within the States'; the Solomon Islands, National Development Plan (1985–9), 'Promote the equitable distribution of the benefits of development'; the Republic of the Marshall Islands, First National Development Plan (rephased for 1986–7 to 1990–1), 'Bring about increasingly equitable development both among different atoll groups of the Republic and among different income groups of the population'; and the Republic of Vanuatu, Second National Development Plan (1987–91), 'Achieve a more even pattern of regional and rural development'. In order to create equitable levels of development throughout the nation, governments focus on the identification of economic activities that the people can pursue in agriculture and fisheries; promotion of traditional skills and crafts and provision of marketing services; provision of support for projects which seek to increase food production for domestic consumption and/or export; and provision of improved transportation and communication links between administrative centres and outer islands.

3. This summary analysis is based on empirical research conducted in several, small, Pacific island countries as part of the United Nations Centre for Regional Development's Research-cum-Training Project on Subnational Development and Planning in Pacific Island Countries.

References

Allen, B. and Hinchliffe, K. (1982) *Planning Policy Analysis and Public Spending: Theory and the Papua New Guinea Practice*, Gower, Aldershot.

Antheaume, B. and Bonnemaison, J. (1988) *Atlas des Iles et Etats du Pacifique Sud*, Gip Reclus/Publisud, Paris.

Berry, R. and Jackson, R. (1981) Inter-provincial inequalities and decentralisation in Papua New Guinea, *Third World Planning Review*, Vol. 3, no. 1, pp. 57–76.

Bertram, I. G. and Watters, R. F. (1985) The MIRAB economy in South Pacific microstates, *Pacific Viewpoint*, Vol. 6, no. 1, pp. 497–519.

Bonnemaison, J. and de Deckker, P. (1990) Cultural identity, development and national integration in Pacific Island nations (paper given at the UNCRD International Conference on Multilevel Development in Pacific Island Economies), Tonga, 10–13 January.

Brookfield, H. and Hart, D. (1971) *Melanesia: A Geographical Interpretation of an Island World*, Methuen, London.

Browne, C. and Scott, D. A. (1989) *Economic Development in Seven Pacific Island Countries*, IMF, Washington, DC.

Chandra, R. and Gunasekera, H. M. (1984) Regional planning and policy in Fiji, in E. B. Prantilla (ed.) *Regional Development: Problems and Policy Responses in Five Asian and Pacific Countries*, UNCRD, Nagoya, pp. 281–339.

Cole, R. V. and Parry, T. G. (eds.) (1986) Selected issues in Pacific Island development, National Centre for Development Studies, *Pacific Policy Papers*, no. 2, Australian National University, Canberra.

Conyers, D. (1981) Decentralization and development: a review of the literature, *Public Administration and Development*, Vol. 4, pp. 187–97.

Corbin, P. (1985) Data requirements for planning in small, Pacific Island countries, in United Nations, op. cit.

Corbin, P. (1987a) Programming and budgeting for subnational development planning in the Pacific Island economies, in P. Corbin, op. cit.

Corbin, P. (1987b) The formulation and integration of subnational development plans in national plans (paper given at the UNCRD Associate Expert Training Meeting on Subnational Planning and Management in Pacific Island Countries), Nagoya, December.

Corbin, P. and Newels, R. (1988) Policy impacts of decentralization measures in Pacific Island economies (paper given at the International Geographical Union, Congress, Sydney) August.

Dommen, E. (ed.) (1980) Islands, *World Development* (special issue), Vol. 8, no. 12, pp. 931–43.

Dommen, E. and Hein, P. (eds.) (1985) *States, Microstates and Islands*, Croom Helm, Beckenham.

Fairbairn, T. I. and Tisdell, C. (1984) Subsistence economies and unsustainable development and trade: some simple theory, *Journal of Development Studies*, Vol. 20, no. 2, pp. 227–42.

Gunasekera, H. M. (1982) Trends in regional planning in the South Pacific, *Regional Development Dialogue* (special issue) pp. 31–49.

Gunasekera, H. M. and Chand, G. (1983) The state of economic development theory in the South Pacific, *Journal of Pacific Studies*, Vol. 9, pp. 218–61.

Gunasekera, H. M. and Lemari, J. W. (1990) Multilevel development planning in the Republic of the Marshall Islands (paper given at the UNCRD International Conference on Multilevel Development in Pacific Island Economies), Tonga, 10–13 January.

Jalan, B. (ed.) (1982) *Problems and Policies in Small Economies*, Croom Helm, Beckenham.

Kuklinski, A. (1980) Comment on Richardson, *Regional Development Dialogue*, Vol. 1, no. 1, pp. 76–7.

Larmour, P. and Qalo, R. (eds.) (1985) *Decentralisation in the South Pacific*, University of the South Pacific, Suva.

Morrison, W. I. (1988) Alternative approaches to modeling subnational development/

planning in Pacific Island economies (paper given at the UNCRD Expert Group Meeting on Development Resource Allocation Methodologies for Subnational Development/ Planning in Pacific Island Economies) 18–23 July, Nagoya.

Newels, R. and Corbin, P. (forthcoming) *A Survey of Institutional Provisions for Subnational Planning in Selected Pacific Island Countries.*

Overton, J. D. (1987) Roads, rice and cane: regional planning and rural development projects in Western Vanua Levu, Fiji, *Pacific Viewpoint*, Vol. 28, no. 1, pp. 40–55.

Paeniu, B. (1989) Interview, *Pacific Islands Monthly*, December.

Premdas, R. R. (1988) Decentralization, development and secession: the case of Papua New Guinea, *Marga*, Vol. 9, nos. 3, 4, pp. 1–15.

Richardson, H. W. (1980) The relevance and applicability of regional economics to developing countries, *Regional Development Dialogue*, Vol. 1, no. 1, pp. 57–79.

Richardson, H. W. and Townroe, P. M. (1986) Regional policies in developing countries, in P. Nijkamp (ed.) *Handbook of Regional and Urban Economics*, Vol. 1, North-Holland, Amsterdam, pp. 647–75.

Selwyn, P. (1975) *Development Policy in Small Countries*, Croom Helm, London.

Srinivasan, T. N. (1986) The costs and benefits of being a small, remote, island, landlocked or mini-state economy, *World Bank Research Observer*, Vol. 1, no. 2, pp. 205–19.

Standingford, J. R. K. (1985) Outer island development policy, *AIDAB/Pacific Regional Team*, bulletin 3, Sydney.

Standish, B. (1979) Provincial government in Papua New Guinea: early lessons from Chimbu, *Monograph 7, Institute of Applied Social and Economic Research*, Port Moresby.

Tabriztchi, S. (1988) Subnational planning models for island economies (paper given at the UNCRD Expert Group Meeting on Development Resource Allocation Methodologies for Subnational Development/Planning in Pacific Island Economies), 18–23 July, Nagoya.

United Nations (1985) *Development Problems and Policy Needs of Small Island Economies*, New York, NY.

United Nations (1987) *Guidelines for Development Planning*, New York, NY.

United Nations Centre for Regional Development (1982) Regional development in island countries, *Regional Development Dialogue* (special issue).

United Nations Centre for Regional Development (1984) *South Pacific: An Annotated Bibliography on Regional Development*, Nagoya.

Ward, R. G. and Ward, M. W. (1980) The rural-urban connection – a missing link in Melanesia, *The Malaysian Journal of Tropical Geography*, Vol. 1, September, pp. 57–63.

World Bank (1988) *World Development Report*, Washington, DC.

PART II:
EXPERIENCES WITH
SELECTIVE INVESTMENT
STRATEGIES

5

DEVELOPMENT POLICY AND ITS EFFECT ON REGIONAL INEQUALITY: THE CASE OF ECUADOR

Stella Lowder

Introduction

Development is a scale- and value ridden phenomenon surrounded by controversy, particularly when the state or international agencies attempt to stimulate it. The rhetoric surrounding development in Latin America and its popular association with improved living standards is matched by criticisms over its impact and the appropriate choice of beneficiaries. These arguments hinge on a number of fundamental principles:

- What role should the state play in development?
- To what extent should efficiency criteria take precedence over equity, or vice versa, in development projects?
- To what extent is it permissible for the financial policies of exogenous institutions and governments to shape the form that development takes within a state (see Simon, this volume).

The importance of these principles has tended to obscure an even more fundamental fact: policies designed to enhance 'productivity' at the national level may have unfortunate consequences for particular subregions and subgroups of society. Few development programmes, with the exception of those for which the *raison d'être* is a specific region, even consider the spatial impact of their policies and their contribution to social inequality. Moreover, development is a process in which all participants exercise political and economic opportunism, and success is generally proportional to the power wielded. The analysis of the effect of development on regional equality in Ecuador presented in this chapter raises issues concerning the state's ability to plan and operate effectively other than at the national level.

The 'proper' role of the state varies according to one's development paradigm. From the 1940s to the 1960s the 'modernization' development model espoused by Latin American governments required rapid industrialization, increased exports and a responsive state prepared to further those ends by public investment in infrastructure and through bestowing fiscal advantages on modern entrepreneurs. Planners, in turn, were much influenced by theories of development that visualized the 'trickle down' of activities from initial growth poles of dynamic industries created in core cities. Efficiency reinforced the concentration of state investment; it was assumed that equity would be achieved over time, as innovations diffused down the settlement hierarchy and across space (Hansen, 1981; Krebs, 1982).

The failure of these policies to engender development outside dominant centres was blamed by supporters on the inadequacies of implementation arising from mismanagement and self-interest. But even during the 1950s, the structuralist school of the UN Economic Commission for Latin America (ECLA) was drawing attention to the exogenous factors that constrained Latin American development (Baer, 1962). Nowadays most analysts would agree that the capitalistic logic of accumulation imposes inequalities both between and within nations (Frank, 1981; Armstrong and McGee, 1985; Portes, 1985). Dependent development at the national level 'structurally polarizes the social order... as the same process dispossesses millions of people from their traditional subsistence without providing industrial employment' (Johnson, 1985, p. 223). Moreover, the action of Latin American ruling groups is better understood if viewed 'as junior partners in the spectrum of financial capitalism' at the global scale (MacEwen, 1986, p. 10), although political considerations may also colour the conditions attached to loans, aid and even export quotas. Furthermore, although the IMF is not a development agency, there is no doubt that its usual recipe for solving balance of payments and liquidity problems, by boosting exports at the expense of wages and the standard of living of the majority, has a substantial impact on development (Edwards, 1988).

Regional equality is also affected by the type of development advocated by the international community. Modern technology stems from a capital intensive world and serves its financial system, which is orchestrated by the international banks. Even the World Bank has been accused of 'using its economic leverage to ensure that they [beneficiaries] accept Western capitalist ideology' (Hart, 1973, p. 208), and its own internal audit department has claimed that 'the Bank was more concerned with lending money, and with the conditions attached to that money, than it was with the quality of its projects' (Hayter and Watson, 1985, p. 160). Thus it is far easier to acquire loans for the construction of large infrastructural projects, or to buy machinery in order to boost exports, than to dispense smaller amounts of credit to domestic food suppliers or for improving feeder roads. Moreover, comparatively small proportions of aid and loans actually reach Latin American shores – 75 per cent of American aid, for instance, is spent on their own consultants, studies and the financing of imported equipment (Cassen *et al.*, 1982, p. 69). These circumstances explain the avidness with which Latin American states accepted loans from the private banks awash with petrodollars in the 1970s;

unfortunately, they also explain the current levels of debt borne by these countries (Thrift and Leyshon, 1988).

The spatial implications of capitalist development are considerable. First, environmental impact studies have been given very low priority. Neither the Alliance for Progress nor the International Bank for Reconstruction and Development incorporated such considerations into their programmes, although they estimated the cost at no more than 3 per cent of the total budget (Pearson and Pryor, 1978, p. 238). Second, administrative structures rarely provided effective counterbalances to the overwhelming demands of the metropolis/metropolises where executives were based. A comprehensive evaluation of Latin American policies and programmes of the mid-1970s concluded that the 'lasting neglect of ... [the decentralization of decision-making] ... gravely prejudices the success of a regional development program' (Stöhr, 1975, p. 120). Third, plans usually portrayed society as a homogeneous population, rather than as a structure sharply defined by gender and social or ethnic status, which condition one's participation in development. Investment is not a neutral factor in development: social attitudes and the structure of society determine how much, for what, where and for whom. The regional loyalties of élites play a major part in shaping the national space economy. The fate of a locality is closely bound up with the power held by its élite and the economic and political opportunities it perceives in development projects.

The case of Ecuador

The goal of Ecuadorian development policies has been to accelerate capitalist production. Ecuador is a small economy by Latin American standards: a great deal of its territory is either Amazonian forest or Andean mountain and its population at the last census (1982) was just over nine million. During the 1960s, indicators suggested that its stagnating banana economy offered few opportunities to undertake regional development programmes (Stöhr, 1975, p. 51). All this changed when its new Amazonian oil fields came on stream just at the time that OPEC forced the price of crude up to unprecedented levels. The international financial community promptly raised Ecuador's credit rating. Furthermore, the state's manoeuvrability was strengthened, as this source of revenue neither threatened domestic vested interests nor demanded structural changes. The solution to all problems was perceived in terms of capital and modern technology; alternative approaches were seldom contemplated. Thus indigenous environmental expertise, including the appreciation of the fragility of Amazonia, was dismissed in the effort to subdue 'natural' constraints, such as drought, floods, erosion and isolation. The contribution of social structures to such phenomena was ignored.

Such attitudes arose from the paradox of a well-funded but weak state: Ecuadorian political parties represent squabbling factions formed around charismatic individuals. Democratic governments have been weak coalitions led either by the sierran élite of Quito, the capital, associated with landlordism, the church and the state, or with that of the largest city, Guayaquil, and its economic interests in agricultural exports, commerce and import-substitution industry. Despite regimes ranging from military dictatorships through liberal to conservative, the image of the state as a mere supporter of private enterprise remained imprinted on society; even measures to encourage investment in specific fields were perceived as infringing managerial freedom (Conaghan, 1988). Personal ambitions and short term objectives rule Ecuador's political life. Paradoxically, such weakness can give rise to entrenched development models, as few dramatic structural changes can survive the almost constant turnover and reorganization of administrators it engenders. The agrarian reform agency, for instance, had thirteen executive directors and several acting directors in the first fourteen years of its existence (FAO, 1980, P. 92).

Centralization, sectoralization and top down policies have been the hallmarks of governments of every political hue. Even if planning is not merely an euphemism for controlling the development efforts of the poor (Black, 1985, p. 529), its repercussions for regional equality are substantial. This is illustrated below with reference to the impact of agricultural and industrial policies; those executed by regional development agencies mostly replicated the dominant development ethos, albeit on a smaller scale. Moreover, the experience of intermediate cities would suggest that seemingly equitable distributional policies cannot begin to redress inherited inequalities. Indeed, one could argue that the physical development of cities has had more to do with the composition of their societies than with the overt policies of the state.

Agrarian development

Ecuadorian agriculture in the 1960s was typified by low yields, precapitalist labour relations and a dominance of minuscule holdings, for 2 per cent of the proprietors controlled 48 per cent of the land (INEC, 1974a). Labour was universally undervalued and wages reflected the assumption that they would be supplemented: on traditional sierra, latifundiae workers had rights to communal pastures, fuel wood, water and a garden plot. Work on coastal plantations was largely seasonal and paid by piece rates; gangs of sierran migrants who left their families at home performed much of it. For decades the sierra had been unable to support its expanding population, and wages for 'free' labour were depressed by the monopolies of the landlords in isolated areas. Three quarters of the landholdings had an area of less than 5 ha in 1974, when activities off the homestead accounted for over 45 per cent of the income of their owners (Chiriboga, 1985, p. 102). Population pressure resulted in the extension of agriculture to fragile and marginal environments, land invasions and accelerated emigration.

Agrarian policy focused on increasing productivity through the capitalization of agriculture, pressurized by need and the modernizing factions of the élite but subject at every turn to the countervailing power of more conservative elements. Most of the tenure legislation, the Agrarian Reforms of 1964 and 1973 and the Law of Agrarian Support and Development of 1979, was passed by military regimes. It took drought, a disastrous rice harvest and increasing land invasions before a civilian government gave guarantees to the coastal sharecroppers in 1970. The legislation put national interests above those of the individual and those of the active before the passive. The 1964 law concerned the expropriation of unworked lands or lands worked by non-waged labour. By the 1960s, isolated estates in marginal environments were already being split up voluntarily by owners unable to extract a profit (Barsky, 1984). Modernizing landlords were happy to settle their redundant labour-force on the marginal peripheral lands of their estates, thus retaining a viable core of the best land and the water rights. If the landowner was not interested in farming, the law stipulated that the lands be divided into viable units and sold to the highest bidder.

The restructuring of land tenure accelerated emigration from some regions. Many proprietors ejected tenants, fearful of them claiming rights under the law. Supposed beneficiaries, especially of landlord-controlled subdivisions, discovered that their new, dry, 5-ha hillside plots were inadequate even for subsistence support. Over the decade after the law, a mere 50,631 beneficiaries received plots, less than 10 per cent of those envisaged (FAO, 1980, p. 93). The second law was more concerned with productivity and introduced the concept of minimum yields to legitimize tenure. In addition, the Empresa Nacional de Abastecimiento y Comercialización (ENAC) encouraged producers by setting prices for selected commodities, while the Empresa Nacional de Productos Vitales (EMPROVIT) distributed cheap, basic commodities to the public. The state subsidized credit via the Banco Nacional de Fomento (BNF). However, the greater capitalization of agriculture meant that casual labour was not needed on the new estates and the displaced rural population had nowhere to go. Consequently, after 1976 the agrarian reform agency devoted nearly all of its energies to colonization. Even so, rural wages declined by 14.7 per cent in real value between 1966–7 and 1977–8 (Tokman, 1981) and the economically active population (EAP) in agriculture dropped by 20 per cent over the same period.

Agrarian policies favoured commercial farmers over small-scale producers owing to their different abilities to gain credit, operate profitably within the fixed-price regime and to produce particular commodities. Gaining credit from the BNF involved such onerous procedures that 72 per cent of the total was gained by less than a quarter of the largest units. Crops destined for the domestic food market tended to be produced on small scale, labour-intensive units that could not cut their production costs. The Ministry of Agriculture reckoned that the area planted with potatoes, barley and kidney beans fell by 37, 76 and 38 per cent respectively over the decade 1970-80, while production declined by 46, 70 and 30 per cent respectively (cited by Lawson, 1988, p. 438). The failure to adjust prices seasonally and to store record harvests played havoc with other annual crops, such as rice, maize and cotton. Meanwhile, subsidies encouraged the expansion of cattle rearing and the import of machinery. Cattle ranching was promoted even in colonization zones, despite requiring extensive pastures since the carrying capacity in many areas was barely 1:1 ha. Industrial crops, which took five or six years to mature, were also encouraged (Lowder, 1982). Most settlers could only clear a few hectares to satisfy their immediate needs; much of their time was necessarily spent raising capital off their holding, often by clearing land for others. The high quality demanded by ENAC and EMPROVIT, their refusal to collect produce and their tendency to pay by cheque and in arrears, resulted in smallholders accepting lower prices from middlemen.

The spatial implications of such policies were embodied in the regional associations with production systems, crops and markets. Small scale production was dominant in the sierra and the pioneer fringe of colonization areas. Most of the potatoes and beans were grown on a small scale in the central sierra provinces, as was the rice in three coastal provinces. Soybeans and cattle were products of large scale commercial enterprises situated on the coast or in Amazonia. The state bias towards large capitalized units resulted in 78 per cent of credit going to the three provinces serving the dominant markets of the primate cities; they represented 80 per cent of the market for liquid milk, for instance (Chiriboga, 1982). The combination of policies benefited the urban consumer, who acquired cheap food at the expense of the producer: average national beef consumption may have increased but not in the countryside.

The state's response to the charge of increasing agrarian inequity took the form of 'integrated rural development projects' designed to incorporate smallholders. Unfortunately, this was seldom achieved. Take for example the project for the Integrated Rural Development of Western Pichincha (IRDWP) costing US $47.2 million in 1980. Three quarters of this sum was contributed by the Inter American Development Bank (IDB). Clearly this forest colonization zone was to be 'integrated' with the national economy rather than internally, given the pattern of investment proposed. The emphasis on construction, which absorbed nearly four times more of the budget than investment in production, did not reflect the priorities of smallholders. But, then, development IDB-style meant that over 80 per cent of the credit was destined to units of 50–99.9 ha, although 87 per cent of the holdings did not reach this threshold. This perhaps explains why the 16.7 per cent of investment earmarked for production was to finance the purchase of cattle, fencing, equipment and to erect sheds on the selected farms, and why only 6.9 per cent of the budget related to contact with potential beneficiaries, in the form of demonstration farms, the rural extension service, technical training for the latter and two 'community development' teams. Even less, 0.7 per cent of the budget, was assigned to the smallholders' first priority, land survey and title registration (Consejo Provincial de Pichincha/IDB, 1980).

This example demonstrates the characteristics of many Ecuadorian development projects, which seldom evaluate the consequences of pursuing modern economic growth. Here, a derisory 2.6 per cent of the budget was allocated to the promotion of forestry, although the *only* activity assessed for productivity, namely cattle ranching, involved clearing it. The impact on one of the most fragile physical environments in the country was ignored, save for a vague research programme aimed at identifying naturally occurring 'economic species' of tree. Some sub-projects were mutually contradictory, such as when a road was located adjacent to an area the Ministry of Agriculture considered of such low potential that it should be left permanently under forest. There was no mention of labour absorption, nor were the views of smallholders obtained. Implementation called for the collaboration of personnel from ten ministries and semi-autonomous agencies not noted for working together. The actual area was selected on political criteria irrelevant to the activities carried out within it: the Ecuadorian sponsor stemmed from a lower tier of government, albeit the council of the capital province, than the institutions it wished to direct. It is not surprising that there is little trace of this project on the ground today (Lowder, 1982).

Despite such shortcomings, Ecuadorian agriculture has been shaped by agrarian development policy: a third of the land under cultivation in 1984 had been affected either by reform (9 per cent) or colonization (24 per cent) (Barsky, 1984, p. 317). Colonization had become the safety valve for displaced rural populations and the area allocated exceeded expectations by 22 per cent. However, only three quarters of the anticipated number of families had been accommodated, owing to land being granted to large commercial enterprises. Participation in schemes was determined by one's capacity 'to formulate demands and present them to the state institutions' whose impact reflected 'whether sectors were or were not capitalized' (Cosse, 1984, p. 91). Cattle ranching expanded and individual yields rose but the production of many food crops declined, such that the index of production for 1983 was only 87 per cent of the 1970 figure. Despite the concern over development, failure to appreciate the implications of measures gave rise to serious problems, which entrepreneurs blamed on state interference with free market forces. ENAC has now been disbanded.

Industrial development

In the 1960s, the industrial sector was dominated by minute units operating in the traditional sectors serving basic needs; almost all manufactures were imported. Successive governments gave industry priority, urged on by the decline in value of Ecuador's commodity exports. ECLA staff assisted in drafting the first Industrial Development Law of 1957. This cast the state as expeditor of supportive legislation, a source of credit and fiscal incentives. The military junta underlined this passivity by promulgating the Finance Companies Act 1963, which concerned the extension of credit to entrepreneurs. The later military regime of 1973–6 directed industrialization via inducements and participated in complementary production itself with the aid of laws promoting small scale industry and crafts, regional industrial zones, industrial parks and tax vouchers. These complemented the national Integral Plan for Change and Development for 1973–7.

These measures were justified by the very low initial base and the need to overcome the obvious constraints of a minute domestic market, low average purchasing power and poor internal communications. Almost every law granted tax relief to industry on a sliding scale according to product priority, innovatory characteristics and location. New, high-priority activities located outside the provinces of the primary cities received the best terms but all industry could avoid 80 per cent of the duties payable on imported machinery, spare parts, technology and on raw materials not available in the country. Many were awarded a five-year tax holiday and dues payable on a range of commercial transactions were quashed. New investment could be discounted against profits for tax purposes by all enterprises (Sepúlveda, 1983). The measures emphasized capital investments, particularly as credit was available at low interest rates and the national currency was so overvalued that these rates were negative through most of the 1970s (World

Bank, 1979). Mechanization proceeded at a great rate to the detriment of the
investment/job-creation ratio. Given the scarcity of technical skills, average factory
wages became among the highest in Latin America, although labour contributed
very little to the costs of production (Vos, 1987). Production increased but the
rate of imports surpassed it two and a half times, while neither employment nor
value added kept pace. Over time the average rate of protection from imported
competition escalated, reaching 135 per cent in 1982 (Fernández, 1983).

The structural impact on the sector was considerable, although change was
initially slow. Landowners and the commercial (import/export) sector were better
placed to provide the collateral demanded. Most of the enterprises that took the
opportunity to modernize catered for the domestic market. The safest subsectors
of brewing, cigarettes and milling imported wheat, attracted some cautious foreign
investment, although enterprises raised most of their capital domestically. By far
the greater part of industry was concerned with consumer non-durables rather than
with the production of intermediate or capital goods. The only activities that could
really be considered to substitute for imports involved TNCs and foreign technology
in the production of construction materials (cement and reinforcing bars), phar-
maceuticals and some metal goods.

Oil money allowed fundamental structural conflicts within the economy to be
temporarily circumvented. The state sought to widen national markets and was
able to subsidize basic consumption while this was taking place. However, the
fiscal mechanisms contributed to the rapid growth of a highly capitalized,
import-dependent sector spatially concentrated in the primary cities: the 20 per
cent differential could not overcome the obvious advantages of market, government
and infrastructure concentrated there. Government credit was biased to the urban
modern sector and three quarters of it went to enterprises located in the provinces
of Guayaquil and Pichincha. Entrepreneurs, protected from outside competition
and operating as oligopolies internally, could reap profits by charging higher prices
to the middle and upper classes rather than by expanding their market to embrace
the poor. Labour absorption was low and on average a third of the installed
capacity was idle. Enterprises expanded mostly horizontally rather than vertically.
The terms of supporting measures never seemed to expire in practice, despite
officially having a limited life. Efforts to foster small scale producers and craftsmen
foundered through their lack of collateral and the capture of the funds by larger
enterprises. Thus the two thirds of the 'industrial' workforce outside the factory
sector gained little from this period, although their markets were threatened by
subsidized produce from the factories. Within the latter, entrepreneurial skill was
concerned more with fiscal dealings than with making the production process
efficient (World Bank, 1984).

Such industry was highly dependent on the state of the world oil and financial
markets. Failure to rectify its shortcomings has been attributed to the uneasy
relationship between the state and the principal beneficiaries, the capitalist
entrepreneurs. Neither military nor civilian regimes won the wholehearted support
of industrialists. Many entrepreneurs stemmed from the commercial sector and
still gained part of their income from imports and exports. They were a highly
inbred community, with 63 per cent of the presidents or managers of the largest

firms being the sons or other close relatives of the founders of the business. Such entrepreneurs interpreted any efforts to direct their actions as an attack on their property (Conaghan, 1988).

Regional development

The inequitable spatial distribution of physical and social infrastructure cannot be accounted for by population differentials alone. Weak regimes would seem to engender top down and aspatial planning that ignores the need to be selective and minimizes formal participation by local authorities. The 1980–4 National Plan, for instance, named all sixteen settlements with populations exceeding 50,000 as growth poles but the mechanics of how they would develop was 'left to negotiation and power struggles between provinces' (Morris, 1981, p. 286). Only some special circumstance has displaced the province as the unit for development planning: four semi-autonomous regional development agencies were created between 1952 and 1971 on these grounds (Figure 5.1). Drought justified the creation of the Centre for Manabí's Recovery (CRM): Manabí suffered from inadequate water to sustain its agriculture, its large rural population and its intermediate cities, Manta and Portoviejo. On the other hand, abundant water, a tropical environment and alluvial soils led an international technical mission to recommend the creation of the Commission for Development of the Guayas Basin (CEDEGE). PREDESUR, the Programme of Studies for Regional Development in the South of Ecuador, arose from an agreement over water use necessitated by the fact that rivers flowed across the border with Peru.

All these agencies interpreted their brief in purely hydrological terms, which increasingly took the form of large scale projects. CRM conceived the Poza Honda Dam and, when this proved totally inadequate, also the Carrizal-Chone project, and more recently, the Daule-Peripa water exchange project. CEDEGE concentrated on schemes in the delta by which excess flood water could be siphoned off to the drier areas to the west. This was also PREDESUR's concern with the Puyango-Tumbes project. These water schemes absorbed enormous sums compared to those invested in regional development *per se*. They undoubtedly increased the supply of water but only to about 2 per cent of the area, comprising the two cities and a small adjacent valley, in CRM's case; CEDEGE and PREDESUR's projects concerned about 3 per cent of their respective areas (Morris, 1981; Pietry-Levy, 1986). About 10 per cent of Manabí's population benefited from the investments made by CRM; PREDESUR's activities assisted only about a fifth of those suffering from drought. Furthermore, not only must a major part of future budgets be earmarked for supportive infrastructure, if the real benefit of these works are to be attained, but also their expense demands that valuable industrial crops utilize the water rather than small-scale producers of beans, rice and maize. This may make sense at national level or even in terms of the resource but it does little to relieve either chronic hardship or massive rural emigration.

CREA, the Centre for Economic Recovery of Azuay, Cañar and Morona Santiago, was created in 1959 to alleviate the deep-seated rural poverty arising from the collapse of the major cottage industry of the area, Panama hats (Espinoza

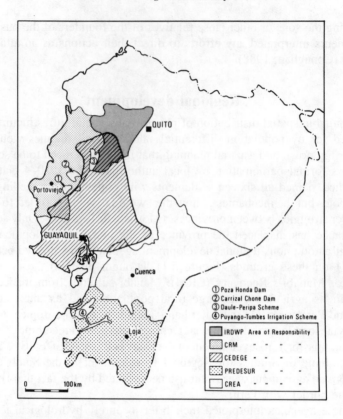

Figure 5.1 Jurisdictions of regional planning agencies in Ecuador

and Achig, 1981). CREA was the only regional development agency to implement widespread small scale investments, which may reflect the size of initial budgets. It constructed rain storage ponds, embankments across gulleys, channels to divert spating streams and dirt roads joining hamlets to village markets. New crafts were promoted along with afforestation schemes, mostly within the vicinity of Cuenca. However the choice of species reflected interest in commercial forestry rather than in the individual peasant or in the fuel supply problems of the area (Morris, 1985). CREA also managed a directed colonization scheme in Morona Santiago. The staff acquired their knowledge of the area through direct contact with the potential beneficiaries and their inclusion in the programmes.

Regimes commonly cite regional investments as proof of their concern but this is no guide to the eventual multiplier effect of the funds. Major construction jobs were undertaken by INHERI, the state irrigation authority, or by international contractors; their decisions were taken in Quito. Agency personnel were based in, and tied to, major cities by low salaries, which ensured that they had to be supplemented by other employment; technicians were thus loath to work in the countryside. Moreover staff, as in most government departments, were either political appointees or young professionals unaccustomed to working with peasants or in difficult environments.

The shortcomings of the management structure bedevilled all the projects. Administrative systems were tortuous: time was wasted waiting for decisions, promised funds and in travelling to the capital for signatures. CREA scored in this respect, as its approach granted it rather more autonomy of action; its own staff executed most of its projects and locals were utilized whenever possible. It also had powerful local champions who were accused of furthering their own interests through its activities, as elsewhere. The Cuenca élite did exert power over a wide area but this does not alter the fact that simple bridges, culverts and dirt roads spread throughout a region benefit a much greater local population than a few showpiece schemes, even if they also allow individuals to gain political capital locally or to benefit from the greater volume of economic activity that results.

Urban repercussions

Rapid capitalist development had substantial impact on the pattern and rate of urbanization. Ecuador's urban hierarchy has long been bicephalous, with Guayaquil and Quito accounting for about two thirds of the urban population. The primary cities attracted 8.5 (Guayaquil) and 7 times (Quito) as many migrants as Cuenca, the third largest city, although they were only 7.9 and 5.7 times as large respectively. There was a marked regional pattern to migration; each primary city dominated its own altitudinal catchment. Significantly, migrants from Cuenca, the only intermediate city for which we have detailed data, changed their focus from the commercial to the political capital during the 'development decade' of the 1970s. As might be expected in a country fragmented by rugged topography, cities developed in each sierra basin and at the coastal termini of routes into the interior. Isolation permitted the growth of such provincial capitals but the agrarian export economy and colonization were responsible for the appearance of new settlements on the coast (Figure 5.2).

There was a marked contrast between the rather subdued intercensal growth rates of the sierra administrative and service centres compared to the new towns of the coast. Milagro, Santo Domingo de los Colorados and Quevedo achieved political recognition as capitals of newly created cantons in 1962, 1974 and 1982 respectively (Table 5.1). Even the older, smaller coastal capitals were transformed in the 1970s, for Esmeraldas became the terminal for Amazonian oil and the site of a refinery, while the environs of Machala proved suitable for the new, more disease resistant Cavendish variety of banana.

Cities benefited disproportionately from the expansion of the bureaucracy. Secondary cities have not been associated with plentiful employment opportunities; in the 1980s, their sex ratios were still female biased while their dependency ratios were much higher than those of the younger rural population. Oil funds allowed a considerable expansion of the bureaucracy and of other labour-intensive services: service employment constituted 8.4 per cent of the urban EAP in 1962 but 22.9 per cent in 1982. However, as resources were distributed according to status in the settlement hierarchy rather than population *per se*, education and health services were invariable concentrated in provincial capitals. These had

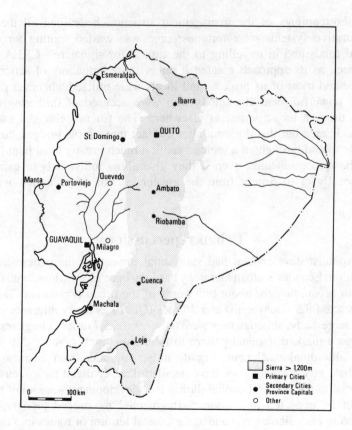

Figure 5.2 Ecuadorian primary and secondary cities

already accumulated a far superior base than the newer cities of inferior status. The state employed 38 per cent of the urban EAP in 1982 but this proportion reached 45 per cent in some of the sierran capitals; the private sector was far more significant in coastal cities.

City residents also gained disproportionately from a series of measures supporting consumption. The greatest subsidies were attached to commodities, such as imported wheat and petrol, which were consumed predominantly there. Subsidized state housing was confined to cities; Guayaquil benefited most but some schemes were located in every major provincial capital. The low interest rates attached to personal credit advantaged educated job holders, who were concentrated in cities. The threshold levels for personal taxation were raised such that barely 10 per cent of the urban population was liable (World Bank, 1979). Such policies did little to distribute income more equitably. Between 1968 and 1975, nearly two thirds of urban income was received by the top 37 per cent of the population in the eleven largest cities, while the share of the poorest quartile was 8 per cent (INEC, 1975). Although average state salaries were not high, they included social security benefits and provided the predictability element required by credit sources, unlike jobs outside the factory or major business sectors. Thus

Table 5.1   Population of primary and secondary cities in Ecuador ('000s)

	1950	1962	1974	1982	Region
Primary cities					
Guayaquil	258.9	510.8	823.2	1,199.3	coast
Quito	224.3	365.7	599.9	866.5	sierra
Secondary cities					
Cuenca	40.0	60.4	104.5	152.4	sierra
Machala	7.5	29.0	69.2	105.5	coast
Portoviejo	29.1	48.3	80.2	102.6	coast
Ambato	34.0	53.4	78.8	100.5†	sierra
Manta	19.0	33.6	64.5	100.3	coast
Esmeraldas	13.2	33.4	60.3	90.4	coast
Milagro	13.7	28.1	53.1	77.0	coast
Riobamba	37.5	41.6	58.1	75.5	sierra
Loja	21.1	26.8	47.7	71.7	sierra
Santo Domingo	1.5*	7.0*	30.5	69.2	coast
Quevedo	4.2	20.6	43.1	67.0	coast
Ibarra	18.1	25.8	41.3	53.4	sierra

Notes
* Estimates by Larrea, 1986, p. 104. Figures for the settlement are not quoted prior to 1974, when the canton of which it is now the capital was created.
† INEC, 1982.

(*Source:* CEDIG, 1986.)

the impact of government expansion and office deconcentration was greater than wage levels implied.

Case study: three cities

The impact of expanding state employment and credit, income concentration and migration varied with the location and the structure of society of the particular city. Isolation can condemn a city to an insignificant economic role or enhance its degree of local autonomy. Likewise, leaders emerge from diverse origins and their interests influence the forms that development takes.

The cities under review here, Cuenca, Ambato and Loja, were all provincial capitals situated in sierran basins. Their potential for integration with the national economy may be gauged from the time taken to reach the primary cities. Cuenca had a daily flight to both cities but public transport took an average of seven hours to Guayaquil and eleven to Quito, assuming that the unsurfaced highways through the complex Andean knot north of the city had not been blocked by landslides or washed out by rains. Loja, to the south of Cuenca, was even more isolated; its airport was reputedly dangerous and roads under construction added nine hours onto the journey from Cuenca, or five to Machala *en route* to Guayaquil.

Ambato's vantage point at the junction of natural routeways was served by mostly paved highways, which sped one to Quito and Guayaquil in two or four hours respectively.

The élites of these cities were formed by different historical processes and were associated with the state to a different degree. Contrast Cuenca, one of the earliest colonial foundations that had administered nearly a third of the country and was associated with many church foundations, their lands and their enterprises, with Ambato, a mere trading post of no account situated at a crossroads. The principal families in the former, despite their modern tastes in consumption, traced their roots back to 'nobles' and associated respectability with the church, the law and the state. Prestige in Ambato was associated with business acumen, much of which was expressed by the intensive, small scale trade of basic commodities within the region. Although Loja was also an early foundation, isolation and border skirmishes limited legal economic opportunities after independence. Local society was very conservative and resisted agrarian reform. Isolation can have its advantages; it was widely believed that much of Ecuador's illicit coca trade passed through Loja.

These contrasts reflected the distribution of resources and the structure of production in their hinterlands. The countryside in Ambato's environs was intensively worked, being composed of irrigated smallholdings the owners of which dominated the national market for potatoes, tomatoes and temperate fruits. This trade contributed to Ambato's commercial prestige and also generated a local market for basic consumer goods at an early stage. Cuenca's élite had severed its ties with agriculture in the vicinity; vegetables were grown intensively on a commercial scale by the descendants of craft workers and traders who had bought land during the Panama hat boom. Elsewhere, poor crops of beans and maize were raised on smallholdings comprising steep and highly eroded fields. Land tenure patterns in the greater region were still very skewed and high productivity, even on the largest units, was rare, although a few of the more accessible estates up valley had invested in dairy herds to supply a milk plant. Pastures were common in the humid environs of Loja, although the peasants, their pigs and fowls were sustained by tiny fields of maize. The more profitable commercial production of sugar cane, peanuts, maize and beef cattle took place at a considerable distance from the city in the valleys to the south and west.

The preference of Cuenca's élite for high status professions and administration was reflected in the city's low industrial base in the 1950s. The industries established from the late 1960s, the tyre plant, furniture works and various processing plants, were established with the aid of subsidies or state capital under the aegis of regional development programmes. They produced mostly final consumer goods utilizing imported inputs, most of which left the region. They had the prestige associated with 'modern' industry and an industrial estate but they did not create much employment. The value added per worker was higher than in Ambato, but the total value of production was only 75 per cent of that of the much smaller city. Ambato's enterprises were small, labour intensive and tucked into niches in the city's fabric; they produced more intermediate goods utilizing a higher proportion of local materials.

Loja's 'industry' remained at the craft level; in the 1980s, the traditional sectors still dominated production in every intermediate city, in terms of both enterprises and employment. Small-scale producers survived; their products were preferred or were cheaper. The shoe and garment industries retained their markets in the face of factory competition from the primary cities by putting out piecework to artisans in surrounding villages. Craft workers also specialized in high-quality production, such as leather goods in Ambato and ceramics in Cuenca. They lost out in mass markets where the factory goods had the aura of status and scale economies granted cost advantages, such as in the case of bottled drinks.

Employment parameters indicate the impact of a decade of development on these cities. The structure of the EAP changed but not necessarily in the manner one might expect from the efforts to promote industry (Table 5.2). Administration exceeded both the commerce and service sectors in employment terms. Moreover, it was the only sector to have expanded significantly more than the total labour force: Cuenca acquired 5,628 new jobs (an 84 per cent increase) within eight years, while Loja and Ambato acquired 3,047 and 2,835 (increases of 112 and 67 per cent) respectively. The proliferation of posts in Loja reflected government fears of a contentious border; most key staff came from other cities. The proportion of 'professionals', i.e. university graduates, in this sector in 1982 ranged from 61 to 68 per cent in Cuenca and Loja respectively.

Table 5.2   Employment by sector in the three secondary cities (%)

	Cuenca	Ambato	Loja
Administration			
EAP 1974	20.38	17.38	21.54
EAP 1982	25.05	21.96	27.62
Commerce			
EAP 1974	15.90	18.22	11.73
EAP 1982	12.91	16.19	9.91
Services			
EAP 1974	16.97	12.60	17.29
EAP 1982	12.02	9.52	11.80
*Production**			
EAP 1974	35.30	37.85	28.19
EAP 1982	36.16	39.98	29.57

Note
* 1974 census categories 'artisans and workmen'; 1982 census categories 7, 8 and 9, which include transport and construction.

(*Source:* Calculations based on the III and IV *Censos de la Población y Vivienda* for 1974 and 1982, INEC, 1974b, 1982.)

The gender ratio became more balanced in all three cities, although women still predominated. The expansion of education was expressed in the decline in the proportion of girls over the age of 12 registered as being 'at home', particularly in Ambato and Loja. But although opportunities improved for the educated woman, they contracted for those seeking productive employment (Table 5.3). The only sector in which gender imbalances declined was that of administrative work. The gender imbalances in unskilled and skilled manual work doubled in the case of Cuenca, where industry was the most modernized, increased by 40 per cent in Loja and by 30 per cent in Ambato. Thus employment for most women became more competitive.

Table 5.3 Women's labour parameters

	Cuenca	Ambato	Loja
Population 1982 ('000s)	152.40	100.50	71.70
Sex ratio 1982 (men per 1,000 women)	868.00	887.00	823.00
% change women 1974–82 (aged 12 and above)			
'At home'	-3.79	-6.63	-5.60
Studying	2.75	2.92	2.03
In EAP	-1.29	1.73	2.14
Men per woman employed 1982			
Total EAP	1.88	2.33	2.28
Manual work	6.02	6.77	12.82
White collar	1.26	1.14	1.32
Commerce	1.25	1.14	1.15
Services	.28	.41	.34

(*Source:* Calculations based on III and IV *Censos de la Población y Vivienda*, 1974 and 1982, INEC, 1974b, 1982.)

The multiplier effect of development policies is difficult to evaluate from such statistics. Less than 60 per cent of the urban EAP were wage earners in 1982, while purchasing power is determined by household income. As family ownership characterized enterprises and savings were discouraged by low interest rates and high inflation, the principal repository for spare cash was property. Thus investment in housing is a useful indirect indicator of increases in disposable income.

All three cities experienced an unprecedented real-estate boom, but the new houses imitated Western styles and were accessible only to the comparatively wealthy. Élites spearheaded the process, by abandoning the increasingly densely populated cores of the cities for peripheral lands, which they developed as exclusive neighbourhoods in which each family designed and built their own mansion. The proportion of wealthy landowning households was limited but the unsatisfied demand for modern housing by the expanding middle-income groups was recognized, for they were able to acquire mortgages for the first time from the rapidly expanding financial sector or the state social security bureau.

The impact of the state's development policies on these cities' fabric can be assessed from the applications processed by the respective planning offices (Table 5.4). These were only set up in the mid-1970s and their documents are incomplete. Even so, the number of plots created was amazing, considering the total number of dwelling units recorded in the cities by the housing census of 1974 (INEC, 1974b). In each city the largest number of projects were undertaken by landlords on a comparatively small scale; over 40 per cent of the land involved privately managed schemes. Co-operatives appeared when middle-income groups, particularly state employees and members of unions, realized that their only hope of acquiring land slightly more cheaply was by purchasing a large plot and subdividing it themselves. Estates arose later, as younger members of the élite became architects and engaged in land development; they began to design standard houses and terraces, subcontract their construction and so produce a home ready for occupancy. It is noticeable that co-operatives were least common in Ambato, where the private sector was more dynamic, while estates were almost non-existent in Loja, where the middle-income groups were primarily state employees not native to the city. The state sector made very little contribution to Cuenca's housing, either because the city was perceived as wealthy enough to provide its own or because the élites orchestrating the process were effective at barring state interference. Certainly, the size and style of Cuenca's new houses were opulent by national criteria (Lowder, 1990a).

Table 5.4   Residential land development, 1970–85

	Cuenca	Ambato	Loja
1982 population ('000s)	152.40	100.50	71.70
Dwelling units			
Number, 1974	18,769.00	15,236.00	8,704.00
Development projects			
Number of projects	497.00	66.00	85.00
Total area (ha)	833.71	257.15	376.80
Number of plots	14,743.00	5,530.00	7,504.00
Developer (% of plots)			
Private landowner	34.22	38.70	34.77
Co-operatives	35.39	16.82	38.33
Estates	26.21	21.08	3.38
The state	4.19	22.68	23.52

(*Source:* III *Censos de la Población y Vivienda* INEC, 1974b; Lowder, 1990b.)

Conclusions

Ecuador's economy boomed in the 1970s as oil prices escalated and oil dominated exports. The state directed development by earmarking oil revenues such that three quarters of its revenues were legally committed before the budget had been drawn up. The country gained pipelines, refineries, hydro-electricity schemes, irrigation and potable water projects, a much expanded electricity grid and highway network, airports and a considerable vehicle stock. Real GDP increased by 8 per cent per annum and average per capita income rose by US $240 between 1972 and 1983 (at 1983 constant prices). Between 1960 and 1980, ten years were added to average life expectancy and there was a 40 per cent drop in infant mortality. When oil prices and output began to fall in 1979, momentum was maintained by external credit. By 1982, the magnitude of Ecuador's external debt frightened lenders and the IMF was summoned to assist.

Although World Bank analysts considered much of the investment worthwhile, they stressed the structural failings of this national-development approach, some of which had been identified in their earlier report (World Bank, 1979, 1984). Fiscal policies had included negative interest rates (in real terms), which had discouraged savings. The overvalued currency had favoured imports over exports, which had covered the former's costs only 71 per cent of the time. The freeze on domestic oil prices between 1972 and 1981 represented a considerable loss of revenue. Industry had grown rapidly but it still contributed less to GDP than elsewhere in Latin America. The relative tax burden had dropped, as had the payments for public services, in a period over which income had grown at unprecedented rates (World Bank, 1984, p. 25).

Such evaluations do not highlight the spatial and social repercussions. Per capita state capital investment in modern urban activities was 11.3 times greater than in the rural, traditional sector in the early 1970s and was still nearly 4 times as great at the end of it (PREALC, 1982, p. 24). Modern production structures spawned and served a domestic demand for income-elastic goods concentrated in the major cities. Agriculture had been forced to subsidize the economy through the main-tenance of high exchange rates and the imposition of prices. While export and industrial crop output increased by 3 per cent per annum, food production declined by the same rate. The unintended regional and social repercussions, especially in the realm of labour absorption, were reflected in mounting migration flows away from Andean districts. Meanwhile the expansion of the bureaucracy and state services was reflected in the growing educated upper-middle-income groups, concentrated in the older provincial capitals, whose tastes in consumption reflected the far wealthier élite preferences. Their demands for improved housing fuelled a real-estate boom that did little for the living conditions of the majority of city residents. Development policies in Ecuador resulted in a superficial modernity that favoured the few in key locations at the expense of the majority.

References

Agosin, M. R. (1979) Analysis of Ecuador's Industrial Development Law, *Journal of Developing Areas*, Vol. 13, no. 3, pp. 263–73.

Armstrong, W. and McGee, T. G. (1985) *Theatres of Accumulation*, Methuen, London.
Baer, W. (1962) The economics of Prebisch and ECLA, *Economic Development and Cultural Change*, Vol. 10, part 2, pp. 169–82.
Barsky, O. (1984) *La Reforma Agraria Ecuatoriana*, Corporación Editora Nacional, Quito.
Barsky, O., Díaz, E., Furche, C. and Mizrahi, R. (1982) *Políticas Agrarias, Colonización y Desarrollo Rural en Ecuador*, CEPLAES, Quito.
Black, J. K. (1985) Ten paradoxes of rural development: an Ecuadorian case study, *Journal of Developing Areas*, Vol. 19, July, pp. 527–56.
Bromley, R. J. (1977) *Development and Planning in Ecuador*, Latin American Publications Fund, Hove, Sussex.
Brownrigg, L. A. (1972) The Nobles of Cuenca: the agrarian élite of South Ecuador (Ph D thesis), Columbia University, New York, NY.
Brownrigg, L. A. (1974) Interest groups in regime changes in Ecuador, *Inter-American Economic Affairs*, Vol. XXVIII, no. 1, pp. 3–18.
Cassen, R., Jolly, R., Sewell, J. and Wood, R. (1982) *Rich Country Interests and Third World Development*, Croom Helm, Beckenham.
CEDIG (1986) *Poblaciones de las Parroquias 1950–1982*, Documento de Investigación, serie Demografía y Geografía de la Población, Quito.
CEPAL (1979) *Ecuador, Desafíos y Logros de la Política Económica en la Fase de Expansión Petrolera*, Cuaderno no. 25, Santiago de Chile.
Chiriboga, M. (1982) La pobresa rural y la producción agropecuaria, in A. Acosta, A. Bocco, M. Chiriboga and I. Fernández *et al.* (eds.) *Ecuador: El Mito del Desarrollo*, El Conejo/ILDIS, Quito, pp. 93–130.
Chiriboga, M. (1985) La crisis agraria en el Ecuador, tendencias y condiciones del reciente proceso, in L. Lefeber (ed.), op. cit., pp. 91–132.
Conaghan, C. M. (1988) *Restructuring Domination: Industrialists and the State in Ecuador*, University of Pittsburg, Pa.
Consejo Provincial de Pichincha/Inter-American Development Bank (1980) *Integrated Rural Development Project for Western Pichincha*, document EC-0117, Washington, DC.
Corkhill, D. (1985) Democratic politics in Ecuador 1979–1984, *Bulletin of Latin American Research*, Vol. 4, no. 2, pp. 63–74.
Cosse, G. (1984) *Estado y Agro en el Ecuador 1960–1980*, Corporación Editora Nacional, Quito.
Ecuador (1980) *Plan Nacional de Desarrollo 1980–84*, Conade, Quito.
Edwards, C. (1988) The debt crises and development: a comparison of major economic theories, *Geoforum*, Vol. 19, no. 1, pp. 3–28.
Espinoza, L. and Achig, L. (1981) *Proceso de Desarrollo de las Provincias de Azuay, Cañar y Morona Santiago*, Don Bosco, Cuenca.
Esser, K. (1985) Modification of the industrial model in Latin America, *Cepal Review*, no. 24, pp. 39–62.
FAO (1980) Country review: Ecuador, *Land Reform*, Vol. 1/2, pp. 89–99.
Fernández, I. (1982) Estado y clase sociales en la década del setenta, in A. Acosta, A. Bocco, M. Chiriboga, I. Fernández *et al.* (eds.) *Ecuador: El Mito del Desarrollo*, El Conejo, Quito, pp. 61–91.
Fernández, I. (1983) Un decenio de industrialización en el Ecuador: un balance crítico, in C. Sepúlveda (ed.), op. cit., pp. 61–134.
Filgueira, C. (1981) Consumption in the new Latin American models, *Cepal Review*, no. 24, pp. 39–62.
FLASCO/CEPLAES (1980) *Ecuador: Cambios en el Agro Serrano*, Quito.
Frank, A. G. (1981) *Crisis in the Third World*, Heinemann, London.
Grindle, M. S. (1986) *State and Countryside: Development Policy and Agrarian Politics in Latin America*, Johns Hopkins University Press, Baltimore, Md.
Hansen, N. M. (1981) Development from above: the centre-down development paradigm, in W. B. Stöhr and D. R. F. Taylor (eds.), op. cit., pp. 15–38.

Hart, J. (1973) *Aid and Liberation: A Socialist Study of Aid Politics*, Victor Gollancz, London.

Hayter, T. and Watson, C. (1985) *Aid: Rhetoric and Reality*, Pluto, London.

Hirschkind, L. (1980) On conforming in Cuenca (Ph D thesis), University of Wisconsin, Madison.

INEC (1974a) *Censo Agropecuario*, Quito.

INEC (1974b) *Censos de la Población y Vivienda*, Quito.

INEC (1975) *Encuesta de Ingresos de Hogares*, Quito.

INEC (1980) *Censos Económicos*, Vol. 7, Quito.

INEC (1982) *Censos de la Población y Vivienda*, Quito.

Johnson, D. L. (1985) Local bourgeoisies, intermediate strata, and hegemony in South America, in D. L. Johnson (ed.) *Middle Classes in Dependent Countries*, Sage, Beverly Hills, Calif., pp. 223–39.

JUNAPLA (1964) *Plan General de Desarrollo Económico y Social 1964–1973*, Book 3, Vol. V, Quito.

JUNAPLA (1972) *Plan Integral de Transformación y Desarrollo 1973–1977*, Editorial Santo Domingo, Quito.

Kasza, G. J. (1981) Regional conflict in Ecuador: Quito and Guayaquil, *Inter-American Economic Affairs*, Vol. 35, no. 2, pp. 3–41.

Krebs, G. (1982) Regional inequalities during the process of national economic development: a critical approach, *Geoforum*, Vol. 13, no. 2, pp. 71–81.

Larrea, O. (1986) Crecimiento urbano y dinámica de las ciudades intermedias en el Ecuador 1950–1982, in F. Carrión (ed.) *El Proceso de la Urbanización en el Ecuador del Siglo XVIII al Siglo XX: antología*, Editorial El Conejo, Quito, pp. 97–132.

Lawson, V. A. (1988) Government policy biases and Ecuadorian agricultural change, *Annals of the Association of American Geographers*, Vol. 78, no. 3, pp. 433–52.

Lefeber, L. (ed.) (1985) *Economía Política del Ecuador: Campo, Región, Nación*, Corporación Editora Nacional, Quito.

Lowder, S. (1982) La colonización como estratégia para el desarrollo: el caso del cantón de Santo Domingo de los Colorados, in R. Ryder and N. Robelly (eds.) *Geografía y Desarrollo*, CEPEIGE, Quito, pp. 127–232.

Lowder, S. (1988) The distributional consequences of nepotism and patron-clientelism: the case of Cuenca, Ecuador, in P. M. Ward (ed.) *Corruption, Development and Inequality: Soft Touch or Hard Graft?*, Routledge, London, pp. 123–142.

Lowder, S. (1990a) Cuenca, Ecuador: planner's dream or speculator's delight?, *Third World Planning Review*, Vol. 12, no. 2, pp. 109–30.

Lowder, S. (1990b) El papel de las ciudades intermedias en el desarrollo regional: una comparación de cuatro ciudades en el Ecuador, *Revista Interamericana de Planificación*, Vol. 23, no. 4, forthcoming.

MacEwen, A. (1986) Latin America: why not default?, *Monthly Review*, Vol. 38, no. 4, pp. 1–13.

Middleton, A. (1981) Class, power and the distribution of credit in Ecuador, *Bulletin, Society for Latin American Studies*, no. 33, pp. 66–100.

Mills, N. D. (1984) *Crisis, Conflicto y Consenso: Ecuador 1979–1984*, Cordes, Quito.

Ministerio de Agricultura y Ganadería (1985) *Economic Perspectives: Grain Pricing Policy in Ecuador*, Quito.

Moncada, J. (1983) *Capitalismo, Burguesía y Crisis en el Ecuador*, Universidad Central del Ecuador, Quito.

Morris, A. S. (1981) Spatial and sectorial bias in regional development: Ecuador, *Tidjschrift voor Economishe en Sociale Geografie*, Vol. 72, no. 5, pp. 279–87.

Morris, A. S. (1985) Forestry and land-use conflicts in Cuenca, Ecuador, *Mountain Research and Development*, Vol. 15, no. 2, pp. 183–96.

Navarro, G. (1976) *La Concentración de Capitales en el Ecuador*, Ediciones Sol y Tierra, Quito.

Pearson, C. and Pryor, A. (1978) *Environment: North and South: an Economic Interpretation*, Wiley, New York, NY.

Pérez Romo-Lerouz, S. (1985) *Crisis Externa y Planificación en Ecuador 1980–1984*, Corporación Editora National, Quito.
Pietry-Levy, A.-L. (1986) *Loja, une Province de l'Equateur*, Centre National de la Recherche Scientifique, Paris.
Portais, M. and León, J. (eds.) (1987) *El Espacio Urbano en el Ecuador: Red Urbana, Región y Crecimiento*, CEDIG, Quito.
Portes, A. (1985) Urbanization, migration and models of development in Latin America, in J. Walton (ed.) *Capital and Labour in the Urbanized World*, Sage, Beverly Hills, Calif., pp. 109–25.
PREALC (1982) *Creación de Empleo y Efecto Redistributivo del Gasto e Inversión Pública: Ecuador 1980–1984*, OIT, Santiago de Chile.
Rauch, J. (1984) An accumulation theory approach to the explanation of regional disparities in underdeveloped countries, *Geoforum*, Vol. 15, no. 2, pp. 209–29.
Sepúlveda, C. (1983) *El Proceso de la Industrialización Ecuatoriano*, Pontifícia Universidad Católica del Ecudar, Quito.
Stöhr, W. B. (1975) *Regional Development Experiences and Prospects in Latin America*, Mouton, Paris.
Stöhr, W. B. and Taylor, D. R. F. (eds.) (1981) *Development from Above or Below?*, Wiley, Chichester.
Thrift, N. and Leyshon, A. (1988) 'The gambling propensity': banks, developing country debt exposures and the new international financial system, *Geoforum*, Vol. 19, no. 1, pp. 55–69.
Tokman, V. E. (1981) The development strategy and employment in the 1980s, *Cepal Review*, no. 15, pp. 133–41.
Townroe, P. M. (1984) The changing economic environment for spatial policies in the Third World, *Geoforum*, Vol. 15, no. 3, pp. 307–33.
Villalobos, F. (1985) Ecuador: industrialización, empleo y distribución del ingreso 1970–1978, in L. Lefeber (ed.), op. cit., pp. 243–91.
Vos, R. (1985) *Government Policies, Inequality and Basic Needs in Ecuador*, working paper no. 22, ISS-PREALC, The Hague.
Vos, R. (1987) *Industrialización, Empleo y Necesidades Básicas en el Ecuador*, Corporación Editora Nacional/FLACSO, Quito.
World Bank (1979) *Ecuador: Development Problems and Prospects*, Washington, DC.
World Bank (1984) *Ecuador: An Agenda for Recovery and Sustained Growth*, Washington, DC.

6

REGIONAL STRATEGY IN COSTA RICA AND ITS IMPACT ON THE NORTHERN REGION

Arie Romein and Jur Schuurman

Introduction

By means of a case study of Costa Rica, this chapter addresses the question of whether there has been a reconsideration of regional policy in those Third World countries that have suffered increasing foreign debt crises since 1980.

In the 1960s and 1970s, Costa Rica borrowed foreign currency to execute an industrialization programme and to diversify its agro-export production. In the 1970s, the country had to face a rise in oil prices and a deterioration of the terms of trade for its agricultural exports. As a result, the country was forced to borrow foreign currency in increasing quantities. Costa Rica's debt amounted to US $1.8 billion in 1980 but was still controllable because of the country's monetary reserves. In the early 1980s however, the Costa Rican economy fell into a severe crisis. The economic growth of the country declined dramatically to a negative level and the debt burden accelerated into an uncontrollable debt crisis: in December 1987, it amounted to US $4.7 billion (DGIS, 1989), a larger per capita debt (US $1,600) than most other Latin American countries.

A useful criterion to evaluate a possible reconsideration of regional policy is the functional-territorial dichotomy (Hinderink and Titus, 1988). In the 1960s, planners and policy-makers in many Third World countries became aware of the growing spatial inequalities, extreme poverty and congestion, resulting from the previous development policy that had been characterized by a concentration of public funds in a few dynamic sectors in their metropolitan areas. Regional policy of a functional type was the first answer: public funds were no longer merely allocated on the basis of national criteria but on regional ones as well. In a functional regional policy, however, these regional development objectives are subordinated to national macro-economic objectives. Regions are conceived of as open systems, the development of which is defined in terms of their functional integration in national economic growth. This means that the resources of the region are directed to serve primarily national development objectives, the regional objectives proper having less priority. The formulation of policies and the decision-making, aimed at functional integration of these regions, occur at the level of the central government.

It has been shown that, in practice, functional regional policy frequently causes regional resources to be exploited indiscriminately without improving regional structures of production (Stöhr and Tödtling, 1978). This has led to the functional paradigm being criticized and the emergence of the territorial paradigm (among others), in which the subordination of regional development to national growth objectives is rejected. A territorial policy would promote the exploitation of regional resources on behalf of the regional population. Agricultural and industrial development in a region should be based as much as possible on local resources, aimed at satisfaction of regional basic needs and relying on regional labour potential. Accumulation of capital and reinvestment should take place in the region itself, which is not regarded as an open system but as one that has to be closed selectively in order to prevent loss ('leakage') of resources (including labour) and capital. For such true regional development to take place, a high degree of regional autonomy in planning and policy formulation on the basis of local decision-making is necessary.

Of course, this dichotomy describes only two positions and does not exhaust all the possible views on regional policy. It can be argued (Gore, 1984) that both paradigms are expressions of the 'spatial separatist theme', in which space is treated as an isolated factor, while the underlying social and economic forces responsible for regional (under-) development are not analysed. This argument is discussed in more detail by Simon (this volume).

In the context of this chapter functionalism and territorialism are appropriate key concepts for analysing the stated problem. If a relationship actually exists between regional policy and debt crisis, any regional policy in Costa Rica will probably have evolved towards an increasingly functional character, for an increasing debt burden would have forced the country to put more emphasis on export production and foreign-currency earnings to pay off its debt (George, 1988) than on a more equal distribution of investment funds over its various regions.

To analyse whether such a possible shift towards a more functional regional policy has occurred, we explore the following two questions: first, are regionally oriented objectives of regional policy (gradually) replaced by nationally oriented ones and, second, has participation by regional bodies in the formulation and implementation of regional policy been reduced or disappeared? If the answers are affirmative, the shift can be said to have taken place.

The effects of the regional policy as implemented will be illustrated with respect to a specific peripheral planning region, Huetar Norte, located in the country's northern lowlands. If the shift towards a functional regional policy has actually occurred, this region's resources should have been allocated to a decreasing number of more profitable economic sectors rather than to the fulfilment of the basic needs of the regional population.

The objectives of regional policy: national or regional?

Until the 1930s, the northern lowlands of Costa Rica, today's Huetar Norte, formed an isolated, sparsely populated and subsistent peripheral region. At that time, Costa Rica faced the disadvantages of its agro-export economy being based on only two products (coffee and, from 1890 onwards, bananas) and on two markets (Great Britain and the USA) (Bulmer-Thomas, 1986, p. 195). Reduced production and exports as a consequence of banana diseases, the First World War in Europe and the Great Depression, resulted in a serious socioeconomic and political crisis. Under these circumstances, the aim of integrating the northern region into commercial markets gained ground. Subsistence agriculture had to be replaced by cash crops for export and food production for the internal market to replace food imports. New export products like sugar cane were stimulated and new laws were passed to increase possibilities for occupying and owning land outside the Central Valley (Figure 6.1). In 1940, the old unpaved road between the Central Valley and the northern lowlands was improved so as to allow year round, motorized transport. The new policy was fairly successful: colonization and commercial production in this region increased considerably.

Following a short civil war in April 1948, the Second Republic was proclaimed. The development policy of the new republic aimed mainly at the reduction of economic vulnerability, a problem inherent in the agro-export model, and at economic growth. Important elements in this new policy were the modernization and diversification of agricultural production, the nationalization of the banking system and the promotion of import substituting industrialization. This last element of the new development model was inspired by an entirely novel approach initiated by ECLA, the United Nations Economic Commission for Latin America. The ECLA approach took as its starting point the shortcomings of a development model based on agricultural exports and promoted a policy shift towards import substituting industrialization and the formation of a Central American common market as the solutions to the economic problems of all the Central American countries. In Costa Rica, the public sector assumed an important role in the policy of economic modernization.

This sector had to be the driving force behind economic modernization and growth through much greater participation in the economy. It has been extended enormously by the allocation of large quantities of financial resources and the creation of many new, semipublic institutions, the so-called '*instituciones autonomas*', in different fields of production, production supporting and social services and public utilities. The number of these semi-state institutions increased rapidly from 13 at the end of the 1950s to 36 at the end of the 1960s and to 60 at the end of the 1970s. The state developed into an '*estado empresario*', a state entrepreneur (Rivera Urrutia, 1982). In 1963, the role of the state in the drive for economic modernization was extended once again. In that year, the first Planning Act in Costa Rican history was promulgated and the state also became a planning institution.

The first two decades of modernization policy did not end economic problems, however. On the contrary, the fluctuations in the price and demand for coffee and bananas continued; industrialization efforts were concentrated entirely in the metropolitan area around San José and were characterized by a high import component; and the growing number of civil servants in San José constituted a heavy burden on the public budget. To compensate for these balance of payments deficits and high public spending, the government saw itself forced to pursue seriously the stimulation of export agriculture once again from the 1960s onwards. The nationalized banking system and the Development Plans were the instruments used to allocate the necessary credits to this sector. The objectives of the First (1965–9) and Second (1969–72) Development Plans were therefore strictly sectoral and were aimed entirely at economic growth by increasing agricultural exports. In the northern region, those Development Plans were directed at the expansion of extensive cattle raising (Table 6.1).

Table 6.1   Land use in the municipalities of San Carlos and Guatuso (%)

Year	Crop cultivation	Grasslands	Not used and other
1955	3.3	10.5	86.2
1973	5.7	34.0	60.3
1984	8.0	67.0	25.0

(*Sources:* IFAM, 1976; DGEC, 1976b, 1987b.)

Another result of the modernization policy pursued after 1948 was increasing socioeconomic inequality between the central region on the one hand and the other regions of the country on the other hand. This problem had become so serious that more attention to the regional implications of the sectoral policies as conducted so far, was inevitable. Against this background, a regional subsystem was included in the 1974–8 Four Year Plan. The main objective of the plan's regional subsystem was the reduction of regional inequalities. This objective, however, was given a 'development accent': the reduction of regional inequalities had to contribute to national economic growth as well. So, although the period 1974–8 marks the birth of explicit regional policy in Costa Rica, its objectives show it to have been of a strictly functional type.

The main measures proposed in the Four Year Plan to reduce regional inequalities were the deconcentration of public institutions from San José to regional urban centres, the development of physical infrastructure in the peripheral regions and the stimulation of agro-industries based on regional raw materials and labour. These objectives were repeated in the next Four Year Plan of 1978–82. The latest innovation in regional development policy is known as '*agricultura de cambio*': agriculture of change. New credits and technical assistance are geared to the promotion of tropical fruits like pineapple, citrus and the less famous maracuya, and some varieties of nuts in the various peripheral regions of the country. Originally intended to improve the position of small farmers in peripheral regions and lessen their dependence on just one or two commercial crops or subsistence production, it now seems to have been incorporated in the overall '*desarrollo de exportaciones*' strategy: the production of non-traditional export crops (Vargas Solís, 1988).

As mentioned before, the Costa Rican economy fell into crisis in the early 1980s. Already in 1979, the government found itself forced to start negotiations with the IMF. Since then, meetings between both parties have taken place almost annually (Schmidt, 1986). To gain the support of the fund in relieving its debt burden, the country was forced to accept the basic prescriptions of this institution's 'structural adjustment' policy: an increase in exports and the reduction of government expenditure.

Both elements have had decisive influences on the country's regional policy. First, it became impossible to realize a substantial and permanent allocation of funds to the erection of agro-industrial plants as well as to the establishment and operationalization of offices for the many public institutions (numbering 93 in 1983) in the regional capitals. Furthermore, the 'agriculture of change' policy has been incorporated into the 1986–90 Four Year Plan, designed entirely to increase export earnings. Although its former aim, improving the position of small farmers, did not exclude production for export, its actual direct orientation to export provides no guarantees to small farmers. In short, the concern with regional inequalities during the 1970s was overtaken in the 1980s by a renewed preoccupation with macro-economic growth, mainly through the promotion of agricultural exports. This becomes even more clear if one considers the regional policy measures actually taken in Huetar Norte in more detail: notwithstanding the pressure put upon the government to reduce spending, the road network and the number of institutions, mainly banks, to support export production have developed impressively in this region (Romein, forthcoming).

In summary, one may conclude that the objectives of development policy have historically always been formulated at the national level to increase national economic growth. In the two Development Plans between 1974 and 1982, regional policy was formally integrated into development policy. But even this regional policy was required to serve a national growth objective as well. Under the pressure of the dramatic increase in the country's foreign debt crisis, the previous accent on sectoral policy, directed to macro-economic growth, has been restored in the last two Development Plans of 1982–6 and 1986–90 and the regional subsystem has received less and less attention.

Implementation of regional policy: co-ordination and the regional framework

In the previous section we mentioned the importance and extensiveness of the public sector in the Costa Rican economy, particularly in the service sector. Nevertheless, private investment makes up the majority of total investment. Furthermore, with regard to regional policy, it should be mentioned that the central planning office, MIDEPLAN (Ministry of Planning and Economic Policy), has neither the means nor authority to determine or to stimulate purposefully the investment and location decisions of the private sector. In general, Costa Rican regional policy is restricted to public investments in peripheral regions, allocated to the establishment of offices and the performing of relevant functions. Any consequential stimulation of private investment in these regions is merely a casual side effect.

A necessary condition for successful implementation of any regional policy under such circumstances is good co-ordination of the planning activities of the various sectoral institutions and departments involved within an adequate regional framework. We now discuss whether this in fact occurs.

One of the aims of the junta which founded the Second Republic was a reduction of state power in its old form, i.e. a state that was controlled by, and dedicated to, the interests of the small coffee oligarchy. Therefore, not only was the public sector extended very rapidly with new, autonomous institutions, but those new institutions were also kept juridically independent of the state. However, the large number of public institutions and the independent status of the majority of them resulted in the dispersion as well as duplication of powers. Autonomous institutions actually outnumbered the state institutions within two decades of 1948. To fight this 'anarchy' in the public sector, several measures were taken in the 1960s and 1970s to lessen the autonomy of the institutions and to improve the possibilities for horizontal co-ordination. Nevertheless, the public sector is still dominated by a conglomerate of numerous more or less autonomous and government institutions, grouped according to their sectoral interests. For example, agricultural production and expansion are covered by IDA (Institute of Agricultural Development), MAG (Ministry of Agriculture and Cattle Breeding), CNP (National Council of Producers) and various product-specific institutions.

As a consequence of this high degree of independence of the autonomous institutions from the central government, MIDEPLAN does not have the legal or the logistical power to ensure co-ordination of their planning activities into an integrated regional plan. In short, co-ordination of the planning activities of sectoral institutions within a regional framework is almost impossible in Costa Rica because of the absence of a suitable institutional framework. Co-ordination could be made possible, e.g. by making the region an administrative unit with its own budget. Since this is not done, regional policy in fact amounts to no more than the implementation of the separate sectoral policies on a regional level (Schuurman, 1987).

Apart from the absence of an institutional framework, the various sectoral institutions that have to co-ordinate their plans also lack a common spatial basis. Since 1915, Costa Rica's administrative structure comprises three spatial levels: provinces, cantons or municipalities, and districts. None of these levels, however, has been assigned a role in the preparation and realization of regional policy. Two reasons can be distinguished: their geographical characteristics, and the lack of political will to decentralize regional policy effectively.

The most disaggregated level, the district, is much too small to serve as a starting point for regional policy. The uniform northern lowlands of the country, for example, comprised about 25 districts in 1973. On the other hand, the provinces certainly are not too small, but their form and location make them inadequate as a basis for regional policy. The northern lowlands, for example, consist of parts of the provinces of Alajuela and Heredia, both characterized by large, centre-periphery contrasts within their borders. The southern parts of both provinces belong to the Central Valley, while their northern parts together make up the northern lowlands. In this way regional inequalities at the national scale, which need to be reduced by regional policy, are reproduced within both provinces. Therefore, they are too heterogeneous to form appropriate entities for regional policy. At the intermediate administrative level, the municipalities are in a better position to function as planning units, although their number might be considered to be too large: the whole country consists of 79 municipalities. Notwithstanding their potential, municipalities play only a marginal role in regional policy because the central institutions lack the political will to supply the municipalities with adequate powers and resources.

If it had converted the municipalities into regional planning entities, the central government would have had to compete with the competence of existing local governments. To avoid this competition, it designed brand new planning regions. In 1975, MIDEPLAN subdivided Costa Rica into six regions, including Huetar Norte. These new planning regions accorded better with the country's spatial diversity, and therefore had greater potential to serve as the spatial basis for regional policy than existing administrative subdivisions. However, this official regionalization has also never fulfilled expectations.

First of all, this regionalization suffered from a lack of continuity, having been modified several times between 1975 and 1988. After the change of Development Plan in 1978, the six planning regions were reduced first to five and later to four. After this last reduction, Huetar Norte was combined with Huetar Atlántico to form the large region, Huetar. During the next plan period, 1982–6, the former region Huetar Norte was restored, but in 1986 the municipality of Horquetas was added to it. In 1988 the same happened with the municipality of Upala (Figure 6.1).

Besides this modification, the official regional division has not been applied by all ministries and public institutions that were required to deconcentrate their activities. In fact, only a small minority of them use this regional classification and most of them have their own regionalizations. For example, DGF (General Directory of Forestry) distinguishes eight regions, as does IMAS (Institute for Social Assistance), but with very different boundaries, while MEP (Ministry of Public Education) works with seventeen regions.

As far as the implementation of regional policy is concerned, one may conclude that both the necessary co-ordination between public institutions and their use of a common spatial basis at the national level are absent. Whether the objectives of the regional subsystem during the 1970s were truly regional or not, the characteristics of the public sector discussed above made possible only the implementation of sectoral development planning (Brügger, 1982).

Figure 6.1 Planning regions in Costa Rica, 1988

Formulation of regional policy: local participation

An important question to be answered in this chapter concerns the extent of local autonomy/participation in the formulation of regional policy. During the 1974–8 planning period, councils for regional policy and regional development at both national and regional levels were added to MIDEPLAN. Both types of council were meant to co-ordinate between sectoral departments and to integrate the planning activities of these departments into one regional plan.

The National Council for Regional and Urban Development (CNDRU) was made up of representatives of several ministries (Planning, Public Works, Economic Affairs and Agriculture) and the presidents of the central bank and various autonomous institutions (water and sewerage, electricity, housing, land reform and municipal promotion). This national council held the final decision-making authority in respect of regional plans, programmes and projects. However, these documents had to be prepared by the regional development councils (CRDs), comprising the presidents of municipal councils, regional citizens' organizations, workers, entrepreneurs and representatives of the regional offices of national public institutions. During the 1978–82 planning period, the competence of the regional development councils was strengthened somewhat and, more importantly from our point of view, subregional development councils (CSDs) were created on a more local scale to represent the interests of local groups in regional policy formulation.

The foundation of CRDs and CSDs could lead one to believe that decision-making in regional policy had acquired an increasingly territorial character. For several reasons, however, this conclusion would be illusory: regional participation in regional policy formulation has in fact always been very small.

First, although represented in the regional development councils, the municipalities were hardly able to defend their own interests in those bodies. Revision in 1970 of the *Código Municipal*, the official document defining the powers, commitments and financial position of the municipalities, gave them many local powers. However, the division of powers between the municipalities and the central institutions was not always clear and, in cases of doubt, central institutions were favoured. With the inception of formal regional policy in 1974, the municipalities' powers were reduced in practice to the administration of such public services as street lighting and refuse collection. More important areas, such as housing and municipal land use planning, are altogether outside their spheres of competence. In the latest Development Plan for Huetar Norte, no important role at all is envisaged for the municipalities. In addition to the absence of real decision-making power in important fields of public policy, the municipalities also lack sufficient funds. In spite of their right to raise and spend taxes, municipal revenues as a proportion of central-government revenues are continually decreasing (Morales, 1986). In 1987 this proportion was reduced to 3.6 per cent (*La Nacion*, 1987), quite insufficient to pay the qualified staff required for policy preparation and implementation.

Just like the municipalities, the representatives of most of the autonomous institutions and government departments in the regional councils could also not make their own, independent decisions. Under the regimes of the 1974–8 and 1978–82 Development Plans, many of these central institutions did deconcentrate their activities by opening regional offices in the 'capitals' of the new planning regions.

The majority of these institutions, however, hardly decentralized any decision-making power to this regional level. The regional offices of the Ministry of Public Health are fairly autonomous, but those of most other public institutions in Costa Rica are virtually limited to offering administrative support for the execution of policy decisions made in the planning departments in San José and aimed at sectoral goals. In sum, despite their legal competence to formulate regional plans, in practice CRDs execute mainly sectoral plans.

Apart from technical and administrative problems, the main reasons for the lack of true decentralization and local participation in regional policy formulation are political. The first one is the reluctance by the administrative élites in the central planning departments to share their power with lower councils. Besides, the political élite in the Central Region is not very amenable to co-operation with autonomous regional councils. This refusal is in fact connected with the composition of the political élite. Both the central government and the Asamblea Legislativa (House of Commons), for example, have always been dominated by ministers and representatives originating from the urban middle class in the 'Gran Area Metropolitana' (Greater Metropolitan Area of San José), with representatives of rural, peripheral areas forming only a small minority. Such an élite is not very interested in minimizing regional inequalities and does not pressurize the reluctant planners to decentralize their power. Finally, there is no political pressure from the region to strengthen its autonomy in policy formulation. During the 1960s and 1970s, the big cattle ranchers of Huetar Norte developed into a regional economic élite. Instead of supporting claims for more autonomy by the regional development councils, it formed an alliance with the central government that bypassed these regional councils. As an important earner of foreign currency, this élite canalized its political power into direct credit lines with the central government and has no interest at all in a regional policy designed to alter the distribution of regional resources and regional credits.

In summary, one may conclude that the formulation and implementation of an integrated regional policy in Huetar Norte have been made impossible by the lack of horizontal co-ordination of the planning activities of sectoral public institutions at the national level, by the lack of continuity and use of a uniform spatial basis and by the absence of true decentralization of policy-making by these institutions. To a large degree, this can be explained by the fact that the political will to implement such a regional policy is absent.

Effects of development policy: the case of Huetar Norte

In the previous sections, some important aspects of the development policy implemented in Costa Rica since the 1960s were discussed. In this final section, the consequences of this policy will be evaluated with reference to one specific region, the Huetar Norte planning region. We focus on the variables of agricultural land use, industrial processing of regional agricultural products and employment.

As mentioned before, this region is located in the northern lowlands of the country, between the Central Cordillera mountain range and the Nicaraguan border. It includes 12 per cent of Costa Rica's 51,100 km^2 and 4 per cent of its 2.4 million inhabitants (DGEC, 1987a). Since the final quarter of the last century it has been a region of spontaneous rural colonization, mainly by people from the densely populated Central Valley. Today the region still has a pronounced rural character, but the social and spatial processes have changed dramatically since development planning has turned it into a beef-exporting region.

The policy to promote cattle raising in the sparsely populated northern lowlands started under the First Development Plan. This policy was quite successful; cattle breeding, in a rather extensive form, has become the dominant land use in the region. The area covered by grasslands in the municipalities of San Carlos and Guatuso, together forming about two thirds of the region, increased from just over 10 per cent to almost 70 per cent in less than thirty years (see Table 6.1). It is quite remarkable that this increase has taken place gradually, but continuously; and independently of the evolution of the regional subsystem in the development policy.

This increase in cattle ranching, to meet the great demand for cheap hamburger meat of regular quality in the USA, was strongly promoted by loans from the World Bank (Keene, 1980). The government invested large sums in the necessary infrastructure, including modern slaughter houses adapted to North American standards, and allocated huge credits to the private cattle ranches in the region. Between 1958 and 1973, the percentage of total agricultural credit allocated to cattle raising in Costa Rica increased from 21 to 60 (Taylor, 1979). Although beef, together with coffee and bananas, was already one of the three main sources of export revenue at the end of the 1970s, the aim of integrating Huetar Norte into the export production of this commodity has been pursued even more keenly since the foreign debt reached crisis proportions after 1980. In the region, efforts are now concentrated more than before on the expansion of extensive and export-oriented cattle raising. In 1985, fully 58 per cent of all bank credits in Huetar Norte, i.e. including not only credits for agriculture but also those for industry, housing, etc., as well, were allocated to cattle raising (MIDEPLAN, 1986).

In 1984, land use in Huetar Norte was distributed quite unevenly between crop cultivation and grasslands, covering 8 per cent and 67 per cent respectively (Table 6.1). The rest of the region's land was (still) covered by forests of various densities. An evaluation of the ecological appropriateness of Huetar Norte for various types of production, published by SEPSA (Central Agriculture Planning Office) and IICA (Inter American Institute for Agricultural Co-operation) in 1985, concludes, however, that 46 per cent of the region's land is adequate or very adequate for crop cultivation (export crops or food crops for the internal market) and 52 per cent suitable for grasslands. A comparison of these figures with the actual land use data demonstrates that the region's ecological suitability for grasslands is overutilized, while the potential for crop production is still barely being exploited. The agricultural potential of the region is being exploited almost exclusively for one product, export beef. Moreover, this underutilization of the ecological potential for crops is much more pronounced in the case of food crops than in the case of such traditional export products as coffee, bananas, cocoa and sugar cane. This means that investments in the region's agricultural sector are not oriented towards the fulfilment of the needs of its own population: the destination of beef and export crops is located outside the region, while current food production is insufficient to satisfy regional demand.

Considering the ecological conditions, expansion of food-crop production would not be problematic, although conditions are not equally suitable for all crops. Expansion would have to take place at the expense of either grasslands or forests. However, reallocation of grasslands to crop production is unlikely because of the importance of beef exports and the political power of the big cattle barons, while continued deforestation is a bleak prospect in a country where the destruction of the tropical rain forests has already acquired the proportions of an environmental disaster. Besides, this deforestation causes severe erosion and largely irreversible soil deterioration. Furthermore, it will also be hard for food crops to compete with new export crops that are being stimulated within the framework of the recent policy of agriculture of change. For example, 27 per cent of the Huetar Norte area is suitable or very suitable for the production of citrus fruits.

The inadequate production of food crops would not be a problem if the actual land use in the region offered enough employment and purchasing power for all its inhabitants to buy food produced outside the region. However, the development policy of the central government means that little use has been made of the labour potential in Huetar Norte.

In the first place, the potential for establishing agricultural processing industries in the region is still hardly utilized, in spite of being an explicit objective of the Development Plans between 1974 and 1982. In 1978, there were six, big meat-processing factories in Costa Rica. None of these was located in Huetar Norte (IFAM, 1979), notwithstanding the fact that 21 per cent of the national cattle stock grazed in this region (in 1984) (DGEC, 1987b). On the other hand, in 1978, 71 per cent of the total slaughter for export was carried out in the Metropolitan Area, in spite of a cattle stock proportion of only 5 per cent in 1984. Apart from these six factories, there are 58 rural abattoirs in Costa Rica, only 5 of which are found in Huetar Norte. In 1984, MIDEPLAN ascertained that most agro-industrial raw materials being produced in Huetar Norte were processed in the Central Region of Costa Rica or abroad. The virtual absence of agro-industry in the region means that, in 1984, this sector offered employment to a mere 5 per cent of the region's economically active population (DGEC, 1987a).

Apart from the paucity of employment in agro-industry, the way in which agricultural credits are concentrated in the extensive cattle-raising sector also does not contribute to job creation. The central government's credit-allocation policy has promoted a highly unequal distribution of investment capital: big cattle-raising landowners receive large amounts of capital to invest in their stock and land, for productive as well as for speculative purposes, while the small landowners have scarcely any capital at their disposal. With their capital, the latifundistas acquire state land (usually virgin forest), but they also buy out small landowners. Therefore regional policy, directed at the expansion of cattle raising, is increasing the inequality of the region's land-tenure structure. The big cattle farms are, however, being exploited very extensively and do not offer enough jobs to employ the growing number of landless labourers and minifundistas. In Huetar Norte, extensive cattle raising requires a labour input of only 6 person days per hectare per year, while the figures for some crops are much higher (Table 6.2) (IFAM, 1976). In other words, more emphasis on crops in regional agricultural policy would create much more employment. At least in this respect, the 'agriculture of change' could make a positive contribution.

Today's reality is that two decades of regional policy, i.e. stimulation of extensive cattle raising, in Huetar Norte have caused increasing migration to the Metropolitan Area, depletion of forest reserves and *'precarismo'* – illegal land occupation. Not only the landless former minifundistas in Huetar Norte itself but also migrants from other regions where similar developments have taken place due to increasing cattle production, e.g. Guanacaste, are unable to find work on the cattle ranches. As a result of this growing presence of 'expelled people' in the region, the northward progress of the colonization front has accelerated, thereby destroying indigenous forest reserves. Possibilities for further land colonization are now virtually exhausted.

The lack of employment opportunities on the latifundios and the end of colonization possibilities have forced more and more people to migrate to the

Table 6.2   Labour input required in average number of person days per hectare per year for various crops or types of land use

Crop/land use	Number of person days
Extensive cattle ranching	6
Maize	38
Rice	49
Sugar cane	76
Oranges	80
Coffee	178
Potatoes	190
Bananas	206

(*Source:* IFAM, 1976, p. 20.)

Gran Area Metropolitana or to illegal occupation of either private land or state property. In the latter case, this means illegal tree felling in order to clear the land for subsistence production. Between 1963 and 1980, 14,000 people were involved in some form of *precarismo* in the two cantons of San Carlos and Sarapiqui alone (Villareal, 1983).

The consequences of development policy in Huetar Norte, therefore, can be summarized as follows: the ecological qualities of the land are utilized almost exclusively for production with destinations outside the region, employment has not kept pace with the demand for jobs, forest reserves are being exploited and congestion problems in the urban areas of the country are worsening.

Summary and conclusions

Costa Rica's national development policy was extended by means of a regional subsystem in the 1974–8 and 1978–82 Four Year Plans. The main objectives of this regional policy were the lessening of regional inequalities and integration of the peripheral regions into national export production – objectives that are not intrinsically incompatible. However, under the influence of the economic crises and the rapidly rising external debt burden in the 1980s, the regional subsystem and the objective of lessening regional inequalities have been suppressed to an increasing degree in the two Development Plans since 1982. Export production to earn foreign currency has again become the main issue in development policy, as it had been in the 1960s.

The main component of development policy in Huetar Norte has always been the stimulation of extensive cattle raising. Only recently has the encouragement of 'non-traditional' export crops, in the context of the 'agriculture of change', been added to this. The policy decisions to activate production of these selected export items in Huetar Norte are being made at the central level. The regional planning councils, created in 1976, are meant to prepare and co-ordinate development planning at the regional level, but their powers are more apparent than real: the real decisions to promote the production of selected export items are being made in San José. The political will of the leading groups in the Metropolitan Area to permit regional policy formulation by more autonomous regional councils, based on the interests of the peripheral regions instead of those of the central regions, is practically absent. Therefore, they do nothing to change this policy formulation 'from above'.

The final conclusion of this contribution is a rejection of the hypothesis that the functional component of regional development policy in Costa Rica has increased in importance under the influence of the foreign debt crisis at the expense of the territorial component. First, development policy has been aimed at earning foreign currency uninterruptedly since the 1960s, albeit for different reasons. In the 1950s and 1960s, balance of payments deficits and high public spending made this necessary, while in the last decade, the high foreign debt forced the country to promote a similar development policy. The example of Huetar Norte demonstrates this very clearly. Even after the integration of regional policy with development policy in the 1970s, when minimization of regional disparities became an important objective, the national growth objective remained intact. Therefore, regional policy, in so far as it was implemented, never had a territorial character and has always been strictly functional. Second, regional pressure groups other than regional élites, which might have designed a regional policy with a territorial character by means of the CRDs, in practice could hardly participate in development-policy formulation. Development policy in Costa Rica has always been very centralized. Third, because of the centralized and unco-ordinated character of development policy, regional policy was dominated by sectoral planning even during the 1970s.

The case of Huetar Norte shows that the impact of the functional regional policy has been negative from the point of view of the majority of the region's population and its natural environment. The fruits of the national development policy in Huetar Norte have been picked by the small cattle-producing élite. True, production figures in Huetar Norte are high due to cattle exports, but at the same time deforestation is continuing and many other inhabitants are unemployed and are being forced into poverty, *precarismo* and migration.

References

Brügger, E. A. (1982) Regional policy in Costa Rica: the problem of implementation, *Geoforum*, Vol. 13, no. 2, pp. 177–92.

Bulmer-Thomas, V. (1986) Central American integration, trade diversification and the world market, in X. Gorostiaga and G. Irving (eds.) *Toward an Alternative for Central America and the Caribbean*, Allen & Unwin, London, pp. 194–213.

DGEC (Dirección General de Estadistica y Censo) (1976a) *Censo de Población 1973*, San José.

DGEC (Dirección General de Estadistica y Censo) (1976b) *Censo Agropecuario 1973*, San José.

DGEC (Dirección General de Estadistica y Censo) (1987a) *Censo de Población 1984*, San José.

DGEC (Dirección General de Estadistica y Censo) (1987b) *Censo Agropecuario 1984*, San José.

DGIS (General Directorate for International Co-operation of The Netherlands) (1989) *Internationale Samenwerking*, March, The Hague.

George, S. (1988) *A Fate Worse than Debt*, Penguin Books, Harmondsworth.

Gore, C. (1984) *Regions in Question*, Methuen, London.

Hall, C. (1985) *Costa Rica: A Geographical Interpretation in Historical Perspective*, Westview Press, Boulder and London.

Hinderink, J. and Titus, M. (1988) Paradigms of regional development and the role of small centres, *Development and Change*, Vol. 19, no. 3, pp. 401–23.

IFAM (National Institute of Municipalities) (1976) *Resumen Cantonal – San Carlos*, San José.

IFAM (National Institute of Municipalities) (1979) *Regionalización de los Mataderos en Costa Rica*, San José.

IICA/SEPSA (Inter American Institute for Agricultural Co-operation/Central Agriculture Planning Office) (1985) *Zonificación Agropecuaria: Esquema Metodológico y su Aplicación al Caso de la Región Huetar Norte*, San José.

Keene, B. (1980) Incursiones del Banco Mundial en Centroamérica, in H. Assmann (ed.) *El Banco Mundial, un Caso de Progresismo Conservador*, DEI, San José, pp. 199–218.

MIDEPLAN (Ministry of Planning and Economic Policy) (1986) *Datos y Cifras de la Región Huetar Norte*, San José.

Morales, M. (1986) Pobreza, participación de la población y costos sociales del crecimiento urbano en ciudades intermedias: los casos de Quesada y Liberia, Costa Rica, in D. Carrion, J. E. Hardoy, H. Herzer and A. Garcia (eds.) *Ciuda-des en Conflicto*, Editorial El Conejo, Quito.

La Nacion (1987), 2 February, Quito.

Rivera Urrutia, E. (1982) *El Fondo Monetario Internacional y Costa Rica, 1978–1982*, DEI, San José.

Romein, A. (forthcoming) *The Rise of a System of Central Places and its Role in Regional Development; The Case of the Northern Frontier Zone of Costa Rica*, Department of Geography, Utrecht.

Schmidt, S. (1986) Die internationale Währungsfonds in Costa Rica: die Strangulierung der nationalen Souveränität, in M. Ernst and S. Schmidt (eds.) *Demokratie in Costa Rica; ein Zentralamerikaner Anachronismus?*, FDCL Verlag, Berlin, pp. 59–84.

Schuurman, J. (1987) *Planificación Regional en Costa Rica y el Caso de la Región Huetar Norte*, Universidad Nacional Heredia, Costa Rica, draft version.

Stöhr, W. and Tödtling, F. (1978) An evaluation of regional policies – experiences in market and mixed economies, in N. M. Hansen (ed.) *Human Settlement Systems: International Perspectives on Structure, Change and Public Policy*, Ballinger, Cambridge, Mass, pp. 85–119.

Taylor, J. E. (1979) Peripheral capitalism and rural-urban migration: a study of population movements in Costa Rica, *Latin American Perspectives*, Vol. VII, no. 25/26, pp. 75–90.

Vargas Solís, L. P. (1988) Vacilante política económica en 1988, *Aportes*, no. 41, February, pp. 11–14.

Villarreal, M. (1983) *El Precarismo Rural en Costa Rica, 1960–1980*, Ed. Papiro, San José.

7

REGIONAL CONSEQUENCES OF OPEN-DOOR DEVELOPMENT STRATEGIES: EXPORT ZONES IN MEXICO AND CHINA[1]

Leslie Sklair

Introduction

Foreign investment has always been widely advertised by the TNCs, who are its main providers, as a resource for development. A new phenomenon of the last decades is that many developing countries now appear to believe that the way forward lies in the promotion of manufactured exports with the help of the TNCs. Even in some developing socialist countries (China, Cuba, Mozambique, for example), the belief in export-led industrialization fuelled by foreign investment and technology (ELIFFIT) is challenging more traditional development strategies, such as import substitution, varying degrees of autarky or exports of primary products. The hope is that earnings from manufactured exports will provide the foreign currency to import what is necessary for successful industrialization. The ELIFFIT strategy locks those who adopt it into the system of global trade.

Despite all the talk of protectionism in both rich and poor countries, the idea of economic autarky has probably never been less popular than it was in the 1980s. No one now appears to believe that the Third World can achieve economic growth, let alone develop, entirely by its own efforts. The experiments in autarky or varying degrees of self-sufficiency popularized by some self-styled communist and socialist developing countries in recent decades have now been abandoned. The reasons for these policy failures are various (see Forbes and Thrift, 1987). Whatever they are, in the 1990s more and more countries, or at least their leaders, appear to see massive foreign inputs as a short-cut to development. These inputs take three main forms, namely, foreign direct investment, aid and loans. Foreign direct investment is the focus here.

We may distinguish between TNCs that invest mainly to gain access to the domestic market (a very common and traditional pattern) and those who invest primarily to assemble or manufacture for export (a less common and more recent pattern). This distinction is not absolute and many TNCs are engaged in both. TNCs are likely to be involved in the export sector in any case.

The rapid growth of export oriented zones (EOZs) in the Third World is a clear manifestation of this phenomenon. Though their total material contribution to Third World economies may be small, EOZs are *symbolic* of an important new development in the global political economy. This can be highlighted by recalling that they were traditionally labelled 'export processing zones'. This nomenclature is becoming increasingly dated as

1. more of the zones are engaging in manufacturing as well as pure assembly activities; and
2. more countries are granting access to their domestic markets for the products of the zones.

In 1970 there were about 20 such zones in ten Third World countries; today there are more than 260 in more than 50 countries (see UNCTC, 1988, pp. 169–72). In most Third World countries the differences between zone and non-zone foreign investment have been declining.

Many have argued that EOZs have failed to transform economic growth into development (notably Frobel, Heinrichs and Kreye, 1980). To evaluate such arguments and the role of export oriented foreign investment in development, a set of criteria for the achievement of positive developmental effects may be elaborated as follows:

1. *Linkages* are the share of imports (backward) and the share of exports (forward) in a firm's products that come from and go to the host economy. The greater the extent of backward linkages (raw materials, components, services) and the greater the extent of forward linkages (sales to intermediate goods industries) achieved with the host economy, then the more likely is the creation of positive developmental effects.
2. The more value added in the host country and the higher the proportion of *foreign currency retained* in the host economy then the more likely the creation of positive developmental effects.
3. The smaller the proportion of expatriate to indigenous managers, technicians and highly *trained personnel*, the more likely is the creation of positive developmental effects.
4. The greater the degree of *genuine technology transfer* (contrasted with technology relocation) the more likely is the creation of positive developmental effects.
5. The more favourable the *conditions of work* are for the labour force (wages, hours worked, job security, workplace facilities) in relation to prevailing conditions in the host society, the more likely is the creation of positive developmental effects.

6. The more *equitable the distribution* of the costs and benefits between the investors, the competing strata among the local populations and the host government, the more likely is the creation of positive developmental effects.

In the worst case, existing linkages might be destroyed, foreign-currency retention might decline, local skills might be lost, the exploitation of labour might intensify and distribution might become more inequitable than before. This means that, however much 'economic growth' (e.g. rising GNP per capita or increasing foreign trade), there is no development. The positive case, where some or all criteria are improving, raises some complex regional questions. It is possible that positive development effects in a region might have negative consequences for surrounding areas. Criticisms of simplistic growth pole theories warn us that we cannot assume that development in one region necessarily leads to all-round development. Singapore's success may have had some negative consequences for the development of Malaysia though it is doubtful that Malaysia would have been better off if Singapore had not been so successful. Singapore has been called an 'overdeveloped zone' and from the point of view of neighbouring regions some EOZs might also appear 'overdeveloped'.

Mexico and China have both utilized export oriented zones as mechanisms to introduce open-door policies in the 1980s and, although the initial conditions were quite different, there are some surprising parallels in the outcomes to date.[2] Mexico is a capitalist society with a substantial public sector that the Mexican state has been privatizing throughout the 1980s. The ruling party, the Institutional Revolutionary Party [in English] (PRI), narrowly won the last election (on official figures that are widely disbelieved) and, though Mexico is not entirely free of repression, it is one of the few Third World countries that approach the criteria of a liberal democracy. The PRI does not entirely dominate civil society. China is a communist society and, although since 1978 the state has been loosening its control over urban and rural enterprises, it is still dominated by centralized planning and the Communist Party. For the Chinese, internal and foreign travel are restricted, there is little freedom to choose one's job and the party controls access to the means of individual advancement. Nevertheless, in both countries open-door development strategies have been promoted vigorously in the 1980s with significant regional and developmental consequences.

The maquilas in Mexico

In 1966 Mexico introduced a Border Industrialization Programme that permitted foreign-owned factories to operate along its border with the USA, duty free. Most of these factories, called maquilas, took advantage of US tariff rules on the re-import of assembled unfinished goods using US-manufactured components.[3]

There are various explanations of why the maquilas were created in the first place, but most agree that the demonstration effect of what are now known as East Asia's 'four little dragons' played a part. The main problem for the Mexican state was to try to play off its heavily protected domestic bourgeoisie against other groups that were already deeply involved with the TNCs. Mexico has always had a reputation as a strict regime for foreign investors but has, for at least a century, hosted vast sums of foreign investment from Europe and the USA. The Border Industrialization Programme, under which the maquilas were introduced, appeared to resolve the dilemma by opening up the remote northern border to virtually unrestricted US investment for export processing, while continuing to protect the rest of the country. Figure 7.1 illustrates the main maquila sites and adjacent US cities.

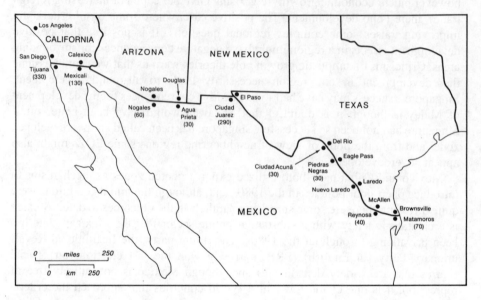

Figure 7.1 Main maquila sites and adjacent US cities (author's estimates of number of plants, 1989, in brackets)

By 1990, Mexico had around 1,500 maquilas, employing 400,000 workers, producing annual export earnings in the region of US $2 billion. The maquila industry is 'booming' along the border and in the interior, where maquilas have been permitted since 1972, but few have ventured. To what extent do the major maquila sites demonstrate positive developmental effects?

Linkages are very meagre, about 1–2 per cent of total purchases. The maquilas make almost no concrete contribution to the growth of domestic industry anywhere in Mexico. The logic of transnational production either encourages, permits or forbids backward linkages. Where a product is entirely horizontally integrated within the TNC or its captive suppliers' networks, or where the intermediate components or materials used are of such a specialized nature that there are simply no available suppliers outside the existing TNC network, then backward linkages are literally forbidden by the logic of global production.[4]

TNCs operating in low cost locations tend to be involved in more traditional and non-state-of-the-art product lines where materials and components are more readily available. This is where, on the surface at least, the logic of TNC production appears to permit backward linkages, and is certainly the case in Mexico where many US-owned maquilas claim to be actively seeking local materials and components. In addition, the Mexican authorities have, for some years, been trying to organize Mexican industry to supply the maquilas with what they need. This is, correctly, seen as a potential bonanza for Mexican domestic industry. The reasons why it has not happened are

1. Mexican production is not of the quality required by the world market;
2. prices are too high; and
3. delivery is unreliable.

No one familiar with the performance of Mexican domestic industry would be amazed by any of this but, even so, one might wonder why the figure for local sourcing is quite so low. The logic of global production is again operating, but in a different way from above. Transfer pricing (where intrafirm transactions are not made at market prices), and captive suppliers (where TNCs are more able to dictate terms), may make it impossible for local suppliers to compete. Only when Mexican factories actually begin to produce what the maquilas need at competitive price, quality and delivery will we be able to prove this one way or the other. There are a few cases where backward linkages have been actively encouraged, where a maquila has gone out to local manufacturers to seek components or materials. Official statistics suggest that maquilas in the interior are more likely to buy local materials than those at the border, which is one reason why the authorities are keen to develop the industry in the interior. The problem for Mexico is that along the border 'local' means both sides of the border and, in some areas (parts of California and Texas in particular), it is not Mexican but US firms that are taking advantage of the opportunities for local suppliers.[5]

In an exclusively export oriented industry, forward linkages with the host economy would not arise. However, it is apparent that the days of pure export processing are numbered precisely because of the ways in which the global economy is evolving. In Mexico, the decisions to allow maquilas to sell to each other, and to allow limited access to the domestic market as a reward for local content, to encourage linkages, are clear signs that export orientation does not exclude both forward and backward linkage effects on the local domestic economy. There is, therefore, nothing in principle to stop maquilas from supplying inter-mediate goods to Mexican producers and there is no doubt that this could be done profitably for both parties. There are already a few cases of US-owned maquilas selling electronic components to Mexican firms. The point here is that it is not simply the ability of the host economy to service linkage requirements that is important, but that the logic of global production may permit or forbid such linkages irrespective of local economic conditions.

The question of foreign-currency earnings, the linchpin of the ELIFFIT strategy, involves the monetary value of the value added in foreign currency due to maquila production. The imports required to produce the exports, the tariffs where applicable and the relative values of the currencies, are all relevant here. Maquila foreign-currency earnings vie with tourism as Mexico's second-place foreign-currency earner behind oil exports. Minimal linkages mean that maquila value added is almost exclusively in terms of wages, utilities and other revenues. The ratio of dollars retained within Mexico varies dramatically with the value of the peso and the spending patterns of the workforce. In 1982, when there were 25 pesos to the dollar, maquila workers were spending more than half of their wages in the USA. In 1989, when there were nearly 2,500 pesos to a dollar, few maquila workers could afford much shopping in the USA (compare, House, 1982, Chapter 7, and Sklair, 1989, pp. 40–1).

There is a good deal of information on the upgrading of Mexican personnel. The author's own research along the border confirms industry claims that Mexican nationals routinely hold positions of managerial and technical responsibility in the maquilas. While most maquilas continue to allocate the very top jobs to US nationals, there are some Mexicans running maquilas themselves. There is no doubt that transnational styles of management and technique are rapidly being transmitted within the maquilas and there is evidence that these styles are permeating through to all levels of Mexican industry. In light of the charge that Mexican domestic industry has been protected for so long that it is unable to meet global competition, the significance of this fact should not be underestimated. One cannot assume, however, that Mexican managers will identify more readily with Mexican national interests and goals, where they conflict with the interests and goals of the corporations that employ them, than foreigners would.

Technology transfer is, apparently, taking place everywhere in Mexico. It is important to distinguish between, for example, General Motors producing auto parts in a General Motors plant (whether in Detroit or in Matamoros) and General Motors producing auto parts in a US plant or in a Mexican maquila. There is a difference between General Motors producing parts totally under General Motors control in any General Motors plant (wherever located) and a Mexican plant genuinely producing auto parts whether for General Motors or anyone else. The former is technology relocation (often misnamed technology transfer), the latter is genuine technology transfer. There is plenty of evidence of technology relocation in the maquilas but very little of technology transfer. However, in some sectors, such as the manufacture of printed circuit boards, and in auto parts, there is certainly more than simple assembly being carried out.

Day-to-day conditions of work reflect the level of exploitation of the workforce. Although all workers under capitalist relations of production are exploited, some workers may be more exploited than others in terms of their respective wages, job security, hours worked and facilities. There is no doubt that maquila workers (and perhaps executives too) are exploited compared with workers (and executives) in comparable US factories – a central reason for the existence of the maquilas. This is one quite valid measure and it is a meaningful interpretation of exploitation. Another is to compare maquila workers with workers in domestically owned Mexican industry, both in the border regions and elsewhere. Maquila wages are low but not the lowest in Mexico (the federal minimum-wage level is higher in the border areas where most of the maquilas are located than elsewhere in Mexico). Job security raises some key issues precisely because the maquilas have often been branded with the label of 'runaway industries', and there have been several notorious cases of US maquilas slipping silently away in the dead of night to avoid paying Mexican workers their legal entitlement of severance pay. During

US recessions, maquilas have closed down (Baird and McCaughan, 1979). Nevertheless, contrary to the poor image of the industry, in the long term maquila plants are not any more likely to close down, industry for industry, than domestic plants in Mexico or the USA. The bare fact is that, due to the maquilas, there are now up to 200,000 new jobs in manufacturing industries in Mexico that did not exist in 1982, when the peso began its decisive decline.

In hours worked there are no substantial differences between the maquilas and the rest of Mexican industry. In fact, the only successful union struggle in Mexico for a 40-hour week took place among the maquila workers of Matamoros in the early 1980s. Workplace facilities, health and safety provisions, subsidized meals, recreation, etc., tend to be a function of plant size rather than ownership. The average number of employees per maquila is over 250, more than double Mexican industry as a whole, which suggests that maquila workers are no worse off than other Mexican workers. This does not mean that all maquilas are healthy and safe places to work, and there are several local studies documenting maquila health and safety hazards (cited in Sklair, 1989, p. 216).

Central to a discussion of the conditions of labour are the questions of worker representation and action. The maquila industry is almost fully unionized in the east (Tamaulipas), somewhat in the centre (Chihuahua) and practically not at all in the west (Baja California). These differences are due mainly to the historical strengths and weaknesses of the unions in the various states. However, labour protest and strikes have tended to be more numerous where unionization is low. Labour turnover is also much higher in the west than in the eastern half of the border, though this is more a function of a tight labour market than tight union control over the workforce (Carrillo and Hernandez, 1985, Chapter IV).

What is certain is that maquila workers, like workers elsewhere in Mexico, have suffered a decline in real standards of living over the past few years while every plunge in the dollar value of the peso appears to herald a new surge in maquila investment. The decline in the dollar value of the peso has hit the maquila worker on the border particularly hard because Mexican border communities have always looked to the USA for some household purchases. Mexican alternative products have often been unavailable to the border consumer, too expensive, of inferior quality or disguised imports from the USA. However, it would not be true to say that the maquila workers are the poorest groups in their localities or nationally. This is a very complex question and it involves not only the maquila worker but her (sometimes his) extended family, often on both sides of the border.

It may be noted that Mexican professionals and the border bourgeoisie have done very well out of the maquilas. Industrial-park development and legal and commercial services for the maquila industry have created a new class of wealthy Mexican maquila facilitators (and, of course, US maquila facilitators as well). The distribution of costs and benefits, unsurprisingly, has favoured these groups over the maquila labour force.[6]

Questions of conditions of labour and distribution in the maquila industry immediately raise further questions of gender – the sexual division of labour. There is now a sizeable literature on 'women in the maquilas' (see, for example, Fernandez-Kelly, 1983, and the chapters by Tiano, Young, Pena and Staudt, in

Ruiz and Tiano, 1987). Maquila employers explain their preference for female workers in the terms that they tend to be more docile and less liable to organize than male workers. It is also significant that the proportion of male workers in the maquilas has been steadily rising over recent years. The sexual division of labour in the maquila industry is in the process of change (compare Fernandez-Kelly, 1983, and Sklair, 1989, Chapter 8).

The growth of the maquila industry can partly be explained in terms of location, though it is also important to realize that many of the US parent corporations of maquilas are domiciled hundreds and sometimes thousands of miles from Mexico. Some of the credit for the success of the maquilas is due to the patterns of regional transnational co-operation that have evolved between US and Mexican private and public agents, devoted both to the economic development of their regions and to private profit. These may be made easier by the border, but they do not depend entirely on it. There have clearly been some positive developmental effects because of the maquila industry, though perhaps not as many as the growth of the industry on the ground would lead one to expect. However, the contribution it has made to Mexican development is much more problematical, particularly as the apparent 'success' of the maquilas has led many Mexican policy-makers to the belief that some version of the ELIFFIT strategy is the way forward for Mexico. It cannot be too strongly emphasized that this 'success' is almost entirely in terms of jobs (very welcome, of course), and annual foreign-currency earnings of between 1 and 2 per cent of Mexico's foreign debt. Nevertheless, the maquila industry is increasingly cited as persuasive evidence that external solutions dictated by the demands of foreign investment and global trade will work for Mexico.

Shenzhen Special Economic Zone[7]

In the late 1970s, as part of the economic reforms that are still sweeping across China, four Special Economic Zones (SEZs) were established, of which the largest by far was at Shenzhen. The regional significance of Shenzhen is obvious. It is contiguous to Hong Kong, which has always been an important market for Chinese primary-product exports and also for trans-shipment trade to and from China. While there was very little infrastructure, practically no industrial base and an almost exclusively peasant labour force, the proximity to Hong Kong and the belief that Hong Kong was a door that could be opened wide on both sides made Shenzhen a viable candidate for regional development. Throughout the 1980s virtually the whole of China's east coast has been opened up to foreign investment and, while Shenzhen was one of the first important sites of foreign investment, it is now one among many. Figure 7.2 illustrates the main locations of foreign direct investment in China.

As in the Mexican case over ten years before, the Chinese had been impressed by the economic results of the 'four little dragons' and were clearly trying to reproduce these in the SEZs. For a faction in the Chinese leadership, this was part of a grand plan to create the conditions for 'China's opening to the world' (Huan, 1986). Throughout its existence Shenzhen, like the open-door policy and liberalization in general, has had its supporters and detractors in different parts

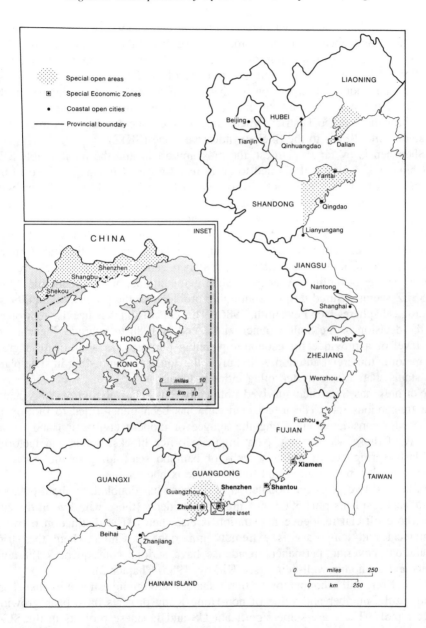

Figure 7.2 Main locations of foreign direct investment in China

of the state apparatus (see Sullivan, 1988).

The Shenzhen SEZ occupies an area of 327.5 km² directly north of the border between Hong Kong (New Territories) and China, acting as a sort of buffer zone between capitalist Hong Kong and communist China. Shenzhen city, ten years ago a village whose only claim to fame was that the Kowloon to Canton railway passed through it, has grown into a sprawling conurbation of over one million inhabitants. Shenzhen's northern border with the rest of the country (Inland China) is also significant for many purposes, such as the control of imports and exports, the flow of foreign currency and the structure of the labour market.

The main industrial expansion has been from the city centre, which is in the middle of the southern flank of the SEZ, west along an industrial corridor that runs through Shangbu District, where most of the zone's electronics industry is located, some 30 km to the Shekou Industrial Zone (SKIZ).[8]

Shenzhen is by far the largest, the most important and the most visible SEZ in China and, at times, it seemed to be in real danger of being sacrificed in the struggle between the 'reformers' who wanted to push ahead at full speed with the open-door policies, and the 'conservatives' who were trying to sustain traditional communist values and slow down, if not actually reverse, some or all of the policies that the open door had let in. This came to a head in 1985 when the then mayor was retired and replaced by a Deputy Secretary General of the State Council (the Chinese 'Cabinet') in order to strengthen central government control on free-wheeling local élites. Attacks on Shenzhen continue, but the economic future of SSEZ seems assured despite continuing problems in the political and cultural-ideological spheres (see Fewsmith, 1986). This conclusion was heavily reinforced by the decision of the State Council at the end of 1988 to upgrade Shenzhen to the level of a province for economic-planning purposes. In addition to a great degree of commercial autonomy, this gives the zone separate listing in state plans – a status shared by only ten other cities in China.

Not only has Shenzhen survived politically, but it has also grown physically at a tremendous rate. The urban landscape has been transformed to the joy of those who consider Hong Kong the apogee of modern civilization and to the dismay of those who do not. New high-rise office blocks, hotels and factories regularly spring out of unlikely webs of bamboo scaffolding, a motorway to Guangzhou has been started and foreign trade is booming.

In the last five years the population of the zone has doubled, to about 600,000 permanent residents plus 500,000 temporary residents (those who live in the city but whose official residence registration is elsewhere). The distinction between permanent and temporary is extremely important in China where the strict regulations covering permanent residence have serious consequences for civil status and economic well-being (see Kirkby, 1985, Chapter 2).

Hong Kong still accounts for the lion's share of foreign investment in Shenzhen, although the number and value of non-Hong Kong projects have been growing quite rapidly. There are some significant US and Japanese projects in the SEZ. Total output, and exports from the zone (around US $3 billion in 1988), both increased more than tenfold between 1984 and 1989 and, despite a recession in 1985–6, the trend is clearly up. Foreign capital 'committed' is around US $4 billion, and the actual foreign investment in operating projects is in the region of US $1½ billion spread over about 1,300 projects. Shenzhen's foreign trade now ranks first in China, having overtaken even Shanghai.

Workers and technicians and managers are flocking into Shenzhen from all over China and business between agencies in the zone and in other parts of China is booming. Economic growth is clearly taking place in Shenzhen – but is it being transformed into development, in terms of the six criteria set out earlier, locally and nationally?

There have been a few notable examples of backward linkages in the SEZ. For example, Chinese flat steel has been used in the manufacture of marine containers (in a joint venture with a Danish firm), and some of the components used in the fast-growing local electronics industry (for example, TV tubes) are produced in China. Some forward linkages have also occurred. The Guangdong Float Glass Company (GFG) (in Shekou), a joint venture of Pittsburgh Plate Glass, Thai investors and a Chinese consortium, produces 130,000 tons of plate glass a year, 60 per cent of which is exported. GFG buys much of its materials in China, including hydrogen and nitrogen piped direct from a nearby supplier, another joint venture, which has captured most of the local industrial-gases market, replacing imports, and has begun to export. The glass company supplies a local mirror factory, completing a fruitful chain of backward and forward linkages.

The key problem here is that China cannot really dictate what foreign investment takes place. Most of the foreign funds that have come into Shenzhen have gone into non-manufacturing sectors (real estate, tourism, infrastructure). In general, Third World authorities will accept practically any foreign investment at all, as the success of policy is invariably measured by the volume of investment attracted rather than its type. In Shenzhen, therefore, industrial linkages have been meagre.

Local value added and foreign-currency retention are bound up with backward and forward linkages, and with the accounting practices of the foreign investor. Details of individual enterprises are rarely available, but the SEZ authorities have released figures on gross output value and profits that indicate something about local value added. Precision is difficult because of the problem of the costing of intermediate goods, particularly for intrafirm transactions, but it is clear that the import bill for SEZ activities far outstrips the export earnings, which suggests that local value added is low. Further, the Chinese have made substantial outlays on infrastructure in the zone and it will be a long time before zone earnings repay these costs.

The third criterion, the upgrading of the local labour force, appears to be very much a reality. There is general agreement between foreign investors and the Labour Services Companies that organize the workforce in Shenzhen that upgrading is taking place, from two directions. First, thousands of technicians and managers from Hong Kong and abroad pass on their skills and, to some extent, their techniques to the local Chinese, at all levels. Second, large numbers of Chinese from other parts of the country have come to Shenzhen to put their skills to use and to upgrade these skills through contact with the new industries. Of course, not all the new industries, not even those with foreign investment, represent technical or managerial advances for the Chinese, but enough of them do to make a difference. The creation of the SEZ has stimulated the establishment of a new university and various colleges geared to the practical needs of the SEZ. Shenzhen's gains in skilled labour may have to be qualified by the losses of the rest of the country, and it remains to be seen whether the benefits outweigh the costs.

Technology transfer is more problematic. On the one hand, the booming electronics industry, particularly the software sector, like joint ventures to create a new generation of Chinese–English electronic typewriters, and Chinese-character operating systems for laser printers, is clearly beneficial to the nascent computer industry in China. On the other hand, there have already been some unfortunate instances in Shenzhen and the rest of China of Western and Japanese firms attempting to pass off obsolete technology and systems to the Chinese, and there is no reason to believe that foreign firms operating in China are any more likely to give up their competitive advantages based on advanced technology than they are in other countries.

The fifth criterion, relating to the day-to-day conditions of work, raises complex problems. In Shenzhen, the traditional 'iron rice bowl', where workers are guaranteed a job for life, is rapidly being replaced with labour contracts where conditions of work and dismissal are laid down (see Leung, 1988). By reforming wage structures and the labour market, the SEZ has attempted to relate effort and performance more directly to reward in ways that appear to contradict Chinese communist theory. Critics argue that, though the iron rice bowl delivers a minimum level of social welfare, it discourages initiative and fails to reward talent.

Covert unemployment, in the form of superfluous personnel, can be found all over China, and SSEZ is no exception. Japanese managers in Shenzhen have complained that because their workers are given such generous leave to visit their homes (30 days per annum for marrieds, 20 for singles), they must employ 10 per cent more labour than they need. It is unlikely that all enterprises in the SEZ do give such generous leave.[9] Nevertheless, gross overmanning is slowly being eradicated in SEZ enterprises. The other side to this question is the exploitation of labour. Paring the labour force down to the absolute minimum often means that demand for increased production has to be met by excessive overtime, which in many cases is either formally compulsory, being embodied in the labour contract, or quasi-compulsory, in the sense that failure to comply results in penalties of various types (withdrawal of bonus, for example). This tends to be part of what we might term 'temporary urbanization' – short term employment of temporary workers who can be laid off when the seasonal demand subsides, and sent back to their places of residence. Therefore they do not represent a continuing drain on urban resources.

It is important to be neither moralistic nor ethnocentric about this. The opportunity to leave home and work for a period in a city for wages that are far in excess of what could be earned at home, irrespective of how they compare internationally, is clearly welcomed by many Chinese, particularly the young. It may also be the case that conditions are no worse than at home, though this is very difficult to ascertain. No doubt some of these temporary 'migrant' workers seek jobs in the SEZ (and in other Chinese cities) unwillingly, but it may not be rural poverty that pushes these people to migrate in search of work so much as the promise of relatively high cash wages and city lights that pulls them. There are some rural poor who come to Shenzhen in search of a better life, such as the pavement shoemenders and hawkers, but they do not tend to be those who work in the factories. As in export zones in other parts of the Third World, foreign investors tend to have the pick of the available working class, and they tend to pick those with at least primary education.

The question of wages is central to this issue. Wage reform has been a contentious issue in China for some years, and the opportunities that the SEZs offered for experimentation in this sphere were enthusiastically seized by cadres and Chinese and foreign managers, often in collaboration. In 1983 all state enterprises in profit began to pay corporate taxes, and their surpluses were available for distribution as the unit decided. Various types of wage-performance linkages were worked out, some quite complex. Wages tend to be split into three parts: basic pay, occupational pay and variable allowance. Comparisons of actual wages in Shenzhen with the rest of China, let alone any other countries, are perilous. The zone authorities fix an official minimum wage and, while temporary workers rarely make much more than this, contract and permanent workers usually do.[10]

One consequence of the combination of contract labour and wage systems that linked rewards to performance was that pressure grew to get rid of unsatisfactory workers, particularly in foreign-invested firms in the SEZ. In 1984, in an interview with the Labour Services Bureau in Shekou, it was reported that about five workers were fired every month, and that most disciplinary requests from managers were acted on (Sklair, 1985, p. 592). The usual reasons for dismissal were unnotified absence from work (3 days or more) and inability to absorb training. Dismissals in Shenzhen were similarly at a very low level. By 1989, dismissals appeared to be much more common, even allowing for the increase in the workforce. Shekou Labour Services Department reported virtually full employment (only 10 out of a labour force of 26,800 were 'waiting for jobs' in January 1989), so the unemployed (for whatever reasons) can usually find jobs very quickly. The same appears to be the case in the rest of Shenzhen. In one factory three permanent workers were fired and subsequently imprisoned for violent behaviour, but were found other jobs by the Labour Services Department when they were released.[11] Unsatisfactory unskilled contract and temporary workers will almost certainly be sent home, but skilled contract and all permanent residents are usually found alternative jobs if they want them.

Two main criticisms of labour conditions have surfaced in Shenzhen, namely child labour and excessive compulsory overtime, both vigorously denied by zone officials and employers. Journalists from Shenzhen, Hong Kong and the USA have reported the employment of child labour in Shenzhen. Under the unambiguous headline 'Shenzhen dismissed more than 500 child labourers', *Shenzhen Daily* (27 August 1988) reported that 529 child workers had been dismissed in Shenzhen. In an attempt to retrieve the situation, the paper argued that 80 per cent of them had lied about their ages. Whatever the truth of particular cases, it is quite likely that children do lie about their ages, encouraged by parents and bureaucrats, in order to get jobs in SEZs (and in the rest of China) that seem so attractive financially to them.

There is general agreement that the workers work very long hours in many foreign-invested factories in Shenzhen. However, there is a profound difference of opinion as to whether this represents exploitation. Management argues that most overtime is voluntary and lucrative, all the more so because it is usually paid in Hong Kong dollars. Critics (some Chinese officials, union leaders, foreign radicals) argue that it is rarely genuinely voluntary, the extra pay does not properly compensate for the extra effort and it masks the fact that capitalists refuse to employ enough workers. The evidence on the question is mixed.

Contract and temporary workers undoubtedly experience very poor living conditions in Shenzhen. Most live in company dormitories, sometimes purpose built by foreign-invested firms but more often rented from various zone authorities. Young women often share four or six to a room and, although rent is usually subsidized by the employer, facilities tend to be very basic.

The other side of the picture is that all permanent and most contract workers in the SEZ have a variety of social insurance packages superior to almost any found elsewhere in China (where the welfare system is a privilege of the small minority who work for the state), covering pensions, medical expenses and unemployment benefit. An interesting feature of these schemes is that they were not won from the employers by a militant labour movement but were imposed on the employers by the local state in Shenzhen and by the China Merchants Company in Shekou. This inevitably raises the question of what the trade unions do. The role of unions has taken some peculiar turns in Shenzhen. Chinese unions since 1949 have tended to perform control and welfare functions for management rather than act as militant representatives of the workers. For example, at least two of the biggest foreign investors in Shenzhen appointed their factory union leaders as personnel managers and this strategy undoubtedly facilitated communications between the union and the management! As usual, there are two sides to the story. Union branches in Shenzhen have to contribute 40 per cent of their revenue to the Shenzhen TU Federation and, as this does not leave enough to pay officials, they usually do it on a part-time basis. It is not certain that union officials on company payrolls stand up for the rights of their workers any less than union officials not on company payrolls in Shenzhen.

Labour relations in the SEZ are, by all accounts, less than entirely harmonious. In 1988, the Shenzhen Labour Bureau reported that, on average, there was a worker–management conflict once every three days in the SEZ. There is a growing realization that under capitalist (or, at least, joint-venture) management, labour disputes and interruptions to production are more likely to increase than to diminish and that the 'cosy' system of labour relations that characterized overstaffing and the iron rice bowl will have to be sacrificed in the quest for reform, particularly in foreign-invested enterprises in the SEZs. The potential for conflict might hinge on the differentiation between permanent and temporary workers. The permanent workforce in Shenzhen is very stable, as is the near-permanent part of the contract workforce, those with skills in high demand who will eventually become permanent workers. Temporary workers and some unskilled contract workers, on the other hand, are in a much more unstable position and a high proportion of labour disputes in SSEZ appears to focus around their demands for better conditions. While the iron rice bowl did not always protect such workers, the combination of labour contracts, rapid growth in the temporary workforce and the foreign investor-driven emphasis on enterprise profits, has altered the balance of power between labour and the state and introduced a new relationship between labour and capital in China.

The sixth criterion, more equitable distribution of costs and benefits, raises some fundamental questions about the structure of Chinese society. While it has rarely been as thoroughly egalitarian as some of its more enthusiastic erstwhile supporters have imagined, in general communist-party policies have managed to keep intraregional inequalities in check more successfully than interregional differences. The economic reforms since the late 1970s, and particularly the exhortations to the workers and the peasants to enrich themselves as fast as possible, within the somewhat elastic bounds of socialist legality, may well have increased local inequalities but, as tends to be the case in China, this depends on the characteristics of the family (age and labour status) and not directly on the ownership of land or property. The benefits of economic growth in Shenzhen as a result of SEZ activity, therefore, are liable to be spread relatively evenly across the mass of the population of the zone and surrounding areas may well share some of these benefits. The question that will need to be answered in Shenzhen, as in the rest of China, is whether the overall benefits of the economic reforms are sufficient to compensate for the income and other differentials they will clearly produce, and for the increasing corruption and nepotism that are widely reported, even in the official media.

A key issue in this regard is the question of whether or not a new class is emerging in China, particularly in and around Shenzhen SEZ. While the assertion that there is a new 'class' emerging in Shenzhen (and in the rest of China) as a consequence of the open-door policy is highly controversial, the assertion that there are new 'strata' (groups that have qualitatively new positions in the social formation) can hardly be denied. There are at least three new strata. First are the *official entrepreneurs*, who deal with foreign investors as representatives of state or parastatal agencies; second are the *private entrepreneurs*, many of whom are in business with foreigners. Often (some might say, too often) these are the same people. Third are the rapidly increasing group of PRC Chinese who live and work in Hong Kong and abroad. These three groups clearly share some interests and have begun to forge some systematic linkages through Shenzhen. One at present unquantifiable but highly significant linkage, common throughout China, is that the officials, private entrepreneurs and expatriate 'red capitalists' are often members of the same family. Children of cadres are commonly engaged in private business and use their parents' connections very profitably, and often corruptly. While in the short term the party and the government may prevent this stratum from becoming a class for itself, as history teaches, new classes have a way of crystallizing.

The other major determinant of social status is gender, as the sexual division of labour in the workplace and women's wages in the domestic economy are also both transformed by the entry of transnational capital. In Shenzhen, unlike most other zones, the numbers of female and male workers are roughly equal and this reflects Chinese employment practices in general, the mediating effects of the labour-supply authorities and the mix of industries. It is also worth noting that the Chinese household, while not by any means free from patriarchal domination, might be somewhat more sexually democratic than households in other Third World countries. However, as both domestic and foreign-invested enterprises become more profit conscious, the relatively generous maternity and other benefits they enjoy are beginning to work against Chinese women.[12] This is not yet a serious problem in Shenzhen, but only because most of the women (temporary or contract) workers are young and would certainly be fired if they became pregnant.

Shenzhen is both a functional and a regional experiment. Functionally, it was created to test the water for the open-door strategy – to give the Chinese some experience in the selective entry of capitalist practices. Regionally, the administration of the SEZ looks forward to 1997 when Hong Kong returns to the PRC. What remains to be seen is whether 'capitalist' production and management techniques can be successfully combined with the virtues of 'socialism with Chinese characteristics' (see Sklair, in Warner, 1987).

Conclusion

In Shenzhen and the maquila sites in Mexico, the logic of transnational production and the precepts of a progressive development strategy conflict at key points. Development, like economic growth, if it is to happen at all, has to happen in particular places at particular times and so development is at the same time a functional and a regional issue. Export oriented zones are one site of the struggle between those promoting the interests of transnational capital and those promoting the interests of the working class. These interests may coincide in the short term, but the evidence from the global system so far (late twentieth century) suggests that they are bound to be fundamentally antagonistic in the long term.

The six criteria for the creation of development effects identify the concrete processes involved in turning economic growth into development. The significance of EOZs, as characterized here, is to acknowledge the ambiguity of a reality in which processes of dependency can be challenged by processes of development. These challenges, however, are not only taking place in abstract theoretical discussion but also in concrete regions, like Shenzhen and all along the USA–Mexican border. It would be unrealistic to expect that those whose interests are served by development strategies that do not deliver the goods to the masses would surrender their privileges without a struggle. Open-door strategies seem to offer a way out of the awful dilemma between dependency without development and capitalist development without social justice but, as the cases of two such different countries as Mexico and China demonstrate, there is little evidence to suggest that this is anything more than a false promise in the interests of transnational capital and its partners, capitalist or otherwise, in the Third World.

Notes

1. I am very grateful to many people for critical discussions on successive versions of this chapter. In addition to the British–Dutch Symposium, I have benefited from lively seminars at the Universities of California (Los Angeles and San Diego), Johns Hopkins, New Mexico, Binghamton, Montreal, Nankai in Tianjin, the Management College in Shekou, the Chinese Academy of Social Sciences in Beijing, Hong Kong University, El Colegio de Mexico in Mexico City, and with my students in development in the sociology department at the London School of Economics.

2. This chapter summarizes part of a wider study to compare the developmental effects of transnational capital in China, Egypt, Ireland and Mexico, provisionally entitled *The Re-formation of Capitalism*.

3. See Carrillo and Hernandez (1985) and Grunwald (in Grunwald and Flamm, 1985, Chapter 4). For the border context, see House (1982) on Juarez–El Paso, and Herzog (1990) on Tijuana–San Diego. What follows is a brief summary of Sklair (1989), which contains a lengthy bibliography.

4. In the author's study of Ireland, where there are many highly specialized TNCs producing for the world market, there are several examples of this (Sklair, 1988).

5. For an interesting framework for the analysis of industrial growth along both sides of the border, see Suarez-Villa (1985).

6. For TNC–Mexican government costs and benefits, see Sklair (1989, Chapter 10), which concludes that the Mexican government could take much more out of maquilas without scaring off the foreign investors.

7. This chapter was written shortly before the Beijing massacre in June 1989. It would be improper to rewrite it in the light of these terrible events, but two brief points may be made. Outrage at the corruption and nepotism referred to in the chapter contributed in large part to the original student protests; and Deng's first public statement after the killings was to reassure foreign investors that the open-door policy would continue.

8. SKIZ is separately managed by China Merchants Holding Company. For general information on the SEZ, see Jao and Leung (1986). The most useful basic sources are *Shenzhen SEZ Yearbook*, *Shenzhen Daily* and *Shekou News*, all in Chinese. I am very grateful to Huang Ping, research student at the London School of Economics, for translations from these publications. Hainan Island has also been given SEZ status.

9. See Sueo (1987). In my own interviews in Shenzhen and Shekou in January 1989 I was told on more than one occasion by Chinese managers in foreign joint-venture companies that their profitability depended largely on being able to dispense with the overmanning typical of most Inland Chinese factories.

10. Differences between the wages of temporary and permanent workers have been cited (see, for example, Leung, 1988, Part Three) as a reason for labour unrest. This was confirmed to the author in an interview with the State Council Special Economic Zones Office in Beijing, in April 1989.

11. Other examples were an electronics assembly company that fires 40 to 50 every year, mainly for not following the factory rules, for poor quality work and for taking days off; another electronics company fires one or two workers a year for being 'a bad influence in society' and has a voluntary turnover of 10–20 workers a year; a biscuit factory where 'labour discipline is generally good and only two or three workers have had to be fired'; and an industrial gases plant where only one worker has been fired, for sleeping on the night shift (interviews in Shekou and Shenzhen, December 1988 and January 1989).

12. See 'Paper urges solving women employment problems', translated in *Foreign Broadcast Information Service* [FBIS]–165 (25 August 1988), pp. 42–3; ' "New thinking" on women's liberation', FBIS-189 (29 September 1988), pp. 51–3; and 'Women struggle for equality', *South China Morning Post* (9 January 1989).

References

Baird, P. and McCaughan, E. (1979) *Beyond the Border*, North American Congress on Latin America, New York, NY.

Carrillo, J. and Hernandez, A. (1985) *Mujeres Fronterizas en la Industria Maquiladora*, SEP/CEFNOMEX, Mexico City.

Fernandez-Kelly, M. P. (1983) *For We are Sold, I and my People: Women and Industry in Mexico's Frontier*, State University of New York Press, Albany, NY.

Fewsmith, J. (1986) Special Economic Zones of the PRC, *Problems of Communism*, Vol. XXXV, December, pp. 78–85.

Forbes, D. and Thrift, N. (eds.) (1987) *The Socialist Third World: Urban Development and Territorial Planning*, Blackwell, Oxford.

Frobel, F., Heinrichs, J. and Kreye, O. (1980) *The New International Division of Labour*, Cambridge University Press.

Grunwald, J. and Flamm, K. (eds.) (1985) *The Global Factory: Foreign Assembly in International Trade*, Brookings, Washington, DC.

Herzog, L. (1990) *Where North Meets South: Cities, Space and Politics on the United States–Mexico Border*, University of Texas Press, Austin, Tex.

House, J. (1982) *Frontier on the Rio Grande: A Political Geography of Development and Social Deprivation*, Oxford University Press.

Huan Guocang (1986) China's opening to the world, *Problems of Communism*, Vol. XXXV, December, pp. 59–77.

Jao, Y. C. and Leung, C. K. (eds.) (1986) *China's Special Economic Zones*, Oxford University Press, Hong Kong.

Kirkby, R. (1985) *Urbanisation in China*, Croom Helm, Beckenham.

Leung W. -y. (1988) *Smashing the Iron Rice Pot: Workers and Unions in China's Market Socialism*, Asia Moniter Resource Center, Hong Kong.

Ruiz, V. and Tiano, S. (eds.) (1987) *Women on the United States–Mexico Border: Responses to Change*, Allen & Unwin, Boston, Mass.

Sklair, L. (1985) Shenzhen: a chinese 'Development Zone' in global perspective, *Development and Change*, Vol. 15, pp. 581–602.

Sklair, L. (1988) Foreign investment and Irish development: a study of the international division of labour in the Midwest Region of Ireland, *Progress in Planning*, Vol. 29, no. 3, pp. 147–216.

Sklair, L. (1989) *Assembling for Development: The Maquila Industry in Mexico and the United States*, Unwin Hyman, Boston, Mass.

Suarez-Villa, L. (1985) Urban growth and manufacturing change in the United States–Mexico borderlands: a conceptual framework and an empirical analysis, *Annals of Regional Science*, Vol. 19, November, pp. 54–108.

Sueo Kojima (1987) Japanese investments in Shenzhen, *China Newsletter*, no. 69, pp. 14–22.

Sullivan, L. (1988) Assault on the reforms: conservative criticism of political and economic liberalization in China, 1985–86, *China Quarterly*, June, pp. 198–222.

UNCTC (1988) *Transnational Corporations in World Development: Trends and Prospects*, New York, NY.

Warner, M. (ed.) (1987) *Management Reforms in China*, Pinter, London.

PART III:
RURAL DEVELOPMENT
AND DECENTRALIZED
STRATEGIES

8

POLICIES AND PREOCCUPATIONS IN RURAL AND REGIONAL DEVELOPMENT PLANNING IN TANZANIA, ZAMBIA AND ZIMBABWE

Carole Rakodi

Introduction

This chapter explores the recent preoccupations of local and regional development and planning in three neighbouring African countries, Tanzania, Zambia and Zimbabwe. The concerns, strategies adopted and policy outcomes are in many respects similar, although there are also important differences. Much agricultural policy has little or no explicitly spatial content, despite the spatially differentiated nature of its outcomes and impact. The agricultural sector has received considerable policy and research attention in all three countries, but space prohibits a review of experience and research findings in this chapter. Instead, this analysis will focus on those components of rural development policy that have had an explicitly spatial component, and on attempts at area-based rural development.

Within subnational divisions, different types of planning, decision-making and implementation processes and procedures have been operated. Understanding of the assumptions on which these have been based is crucial to analyses of their impact. A common theme here is the influence of donor agencies. Finally, national urban development strategies have been important on the agenda of all three countries, informed by perceptions that the cities are growing too fast, becoming too large and absorbing a disproportionate share of resources, and that smaller settlements have a role to play in regional development. The assumptions on which the policies have been based and their outcomes are assessed for the countries in turn.

Agricultural and rural development

Tanzania and Zambia have faced similar difficulties in developing agricultural production for both domestic use and export since gaining independence in the 1960s. They have followed similar paths and have experienced similar difficulties. Despite an ideological commitment to lessening regional disparities and assisting peasant agriculture, policy-makers have tended to see large scale agriculture (both private and public) as the easier option, despite much evidence to the contrary. This policy bias has been exacerbated by the lack of understanding of, and respect for, peasants and their agricultural systems; the realities of political power distribution, especially in Zambia; and the need for foreign exchange, which has biased policies against food production for domestic use.

Public-sector efforts to control prices and marketing have attempted to achieve the conflicting objectives of keeping urban food prices down, extracting revenue and increasing production and revealed a gap between policy aims and administrative capacity. Failure of the state to manage the system, exacerbated by poor world prices for exports and restricted supplies of inputs and consumer goods, led to the withdrawal of peasants into subsistence production or the parallel economy, the latter particularly in Tanzania. Many of the agricultural and rural development policies adopted have been ill thought out, *ad hoc* and single prong. Moreover, attempts to increase production have been hindered by the undoubted physical constraints (Meyns, 1984; Good, 1986a, 1986b; Kydd, 1986; Makgetla, 1986; Raikes, 1986; Svendsen, 1986; Bryceson, 1988; Burdette, 1988; Colclough, 1988; International Labour Office, 1988).

Zimbabwe's commercial agricultural sector is better developed than in either Tanzania or Zambia, although dependence on it for food and export crop production has ensured that it continues to consume a disproportionate share of credit and agricultural services (Weiner, 1988; Stoneman and Cliffe, 1989). Nevertheless, dramatic increases in yields and volume of production have been achieved since independence in 1980 by the peasant sector in the communal areas. This positive response on the part of farmers with access to productive resources, including land, credit, off-farm income and reliable rainfall, has, however, resulted in increased inequalities within the areas (Mumbengegwi, 1986; Cliffe, 1988a, 1988b; Jackson and Collier, 1988).

In all three countries, the increases in producer prices during the 1980s, while undoubtedly beneficial to producers, showed that demand-side policy will have only limited and uneven effects while supply constraints and inequalities in access to factors of production remain. Policies, while ostensibly aimed at reducing regional disparities, were rarely placed in a spatial framework. Exceptions are the villagization and settlement policies adopted to varying extents in the three countries, and these will now be discussed in turn.

Villagization in Tanzania

In Tanzania, changes to the organization of agricultural production in order to foster the production of export crops, in particular, have been associated with settlement policies. Both were centrally formulated and implemented. The aims of *ujamaa* were to relocate the dispersed rural population in nucleated settlements and to collectivize farming. Between 1967 and 1970, the programme was implemented on a voluntary basis. However, where communal agriculture and enterprises were introduced, these tended to be poorly administered and gave rise to conflicts (Brebner and Briggs, 1982; Kauzeni, 1988). Impatience with the slow progress led to the adoption, in 1973, of the villagization programme, with its twin aims of rationalizing, although not collectivizing, farming and facilitating the delivery of production and social services by concentrating population and providing infrastructure. This programme had a much wider impact. While in 1973 it was estimated that only 2 million rural Tanzanians lived in villages, by 1977 this had increased to 13 million (Legum, 1988). This programme, together with increases in the expenditure on social services, has resulted in increased access, especially to primary education, health care and co-operative retail facilities, on a relatively equitable basis (Maro and Mlay, 1979; Kulaba, 1982; Kauzeni, 1988; Legum, 1988).

In the more densely settled commercial agricultural areas, the need for population relocation was limited. However, in much of the country, large scale relocation had a temporary disruptive effect on production (Brebner and Briggs, 1982; Raikes, 1986). A significant number of villages were poorly sited in relation to soils and water (Kulaba, 1982) and, although the intention was to provide a water supply for every village, lack of funds for transport, construction and maintenance prevented the realization of this aim (Therkildsen, 1986). The concentration of population led to deforestation and to a tendency to cultivate fields near the village on a long term basis, with detrimental effects on soil fertility (Nilsson, 1986; Raikes, 1986).

While provision of services has increased basic needs satisfaction, it has also increased dependence on the state and foreign aid (McCall and Skutsch, 1983) and given rise to 'a powerful interest group within the peasantry which implements, benefits from and sometimes depends upon the system of state, party and parastatal controls over the flow of materials and over how peasants [villagers] use their time' (Raikes, 1986, p. 126). Bureaucratic delivery of services has come to be characterized by inefficiency, negligence and often corruption. Bryceson (1988) attributes this to the coincidence of the villagization programme with a deterioration in the wages of state employees, which resulted in a tendency to indulge in clientage practices, but outside the kinship network that had checked the activities of earlier patrons. The loss of control over their livelihoods to the civil service and the party gave rise to considerable resentment on the part of peasants, although analysts differ on the extent to which they think this extension of control was an actual aim, albeit implicit, of the policy (Brebner and Briggs, 1982; Bryceson, 1988; O'Connor, 1988).

Although regrouping of villages to facilitate the provision of services has been attempted on a small scale in Zambia, it has been neither popular nor successful. Attempts at villagization have been made in Zimbabwe, particularly along the Mozambique border, but people are resistant to moving unless services are provided first, and many have unpleasant memories of being herded into 'Protected Villages' during the liberation struggle.

Settlement schemes in Zimbabwe

Settlement schemes have formed a further element in agricultural policy, of greatest significance in Zimbabwe, although even here less extensive than originally intended (Cliffe, 1988b). Almost 3 million ha of large farms (5 per cent of the land area) have been acquired since 1980, on which about 40,000 families have been resettled, most under a system whereby households receive 5 or 6 ha of land for individual cultivation in addition to access to common land for grazing, and a few under co-operative arrangements. Land in resettlement schemes is allocated to families. Women are excluded from access to land in their own right, with the exception of widows. Resettlement has, therefore, perpetuated women's customary lack of rights to land (Jacobs, 1984).

Much of the land on which resettlement schemes have been established was previously under-utilized and, once the schemes are under way, they are proving comparatively productive in the better agro-ecological zones, although their full potential has not yet been realized (Cliffe, 1988a, 1988b; Weiner, 1988; Stoneman and Cliffe, 1989; Wekwete, 1989b). The planning and development of the schemes are largely a central-government function, resulting in a lack of integration between schemes and their surrounding areas in productive, infrastructure and administrative terms, a problem exacerbated by their scattered distribution (Wekwete, 1989b). Dissatisfaction with the progress of the resettlement programme, which has slowed even further since the mid-1980s, has led to widespread squatting on vacant or unused land. Squatters have been supported by ZANU and opposed by the large scale commercial farming sector, resulting in an ambivalent government response as it tries not to alienate either of these power groups (Jacobs, 1984). Thus, although resettlement policy has been underpinned by an ideological commitment to redressing the inequalities in land holding that were a major impetus for the liberation struggle, the need to maintain production by the large scale sector and the resource costs of the schemes have limited its scope. Further opposition to land redistribution may be expected, especially from large scale commercial farmers as well as from agricultural bureaucrats (Moyo, 1986; Cliffe, 1988b).

Differences in the relative priority given to villagization or settlement schemes in the three countries reflected differences in ideology and priorities rather than in patterns of land ownership or conditions in the more sparsely settled rural areas. Policies were mostly centrally planned, despite the varying conditions of the regions in which they were implemented. Often the relationship between settlement schemes or state farms and their surrounding regions has been neglected in both planning and assessments of policy outcomes.

Administrative decentralization

Attempts to develop rural areas have been closely linked to administrative decentralization, with the twin aims of improving policy formulation and implementation and increasing participation in decision-making and development. The decentralized systems of government adopted in the three countries will be reviewed here, from the point of view of administrative effectiveness and political acceptability. The area-based approaches to regional development that have been adopted and their outcomes are closely related to the administrative structure and will be analysed in the subsequent section.

Tanzania

In Tanzania, local administration during the 1960s was by District Councils, but these were weak and poorly funded. Their ineffectiveness was compounded by increasing demands, unmatched by resources, made upon them by central government (Mawhood, 1983). All subregional planning was sectoral (Mosha, 1988). The changes in policy emphasis following the Arusha Declaration led to a search for ways of increasing participation.

In 1972, a hierarchical administrative system was established, in which the districts were amalgamated into twenty regions (see Figure 8.1), each with a political commissioner and a Regional Development Committee made up of representatives and officials and responsible for regional policy formulation (Mawhood, 1983). In a bid to reduce interregional differentials, 40 per cent of the national development budget was to be allocated to the regions. Districts were made very much subordinate, although District Development Councils and their Planning Committees were supposed to formulate and implement plans (Kauzeni, 1988). Regional Integrated Development Plans (RIDEPs) were prepared for each region between 1974 and 1976 and will be discussed later. However, regions had no independent source of revenue. Instead, they were dependent on inadequate allocations of funds from central government (a peak of 17.2 per cent of the national development budget in 1975–6 and roughly 25 per cent of the recurrent budget) and their ability to attract donors. Despite the supervisory responsibility for regional development allocated to the Prime Minister's Office, co-ordination between public agencies at national and regional levels has remained poor, while some local authorities have rejected proposals contained in the regional plans (Mosha, 1988; Mtui, 1988; Ngasongwa, 1988). In 1975 a village-level administrative system, comprising Village Assemblies, elected Village Councils, and their constituent committees, was established and reinforced in 1982. These were intended to formulate and execute policy respectively and to present proposals to the district for approval and funding (Asmerom, 1984).

Views on the extent of participation achieved as a result of decentralization in Tanzania vary. Freedman (1985) is generally positive, while Samoff (1979), Mawhood (1983), Kauzeni (1988) and Mosha (1988) are more critical. Mawhood dismisses regionalization as deconcentration with advisory committees. Mosha's explanation is largely based on the deficiencies of the administrative structure and planning process. Samoff notes the authoritarian nature of Tanzanian administration (although occasionally local areas avoid and subvert central government directives) and suggests that government expects citizens to approve and legitimize government actions rather than take an active role in decision-making. The technocratic orientation of the new institutions is related to the nature and ideology of the governing class, the bureaucratic bourgeoisie, which has been able to resist attempts by progressive elements at a national level to provide subordinate classes with

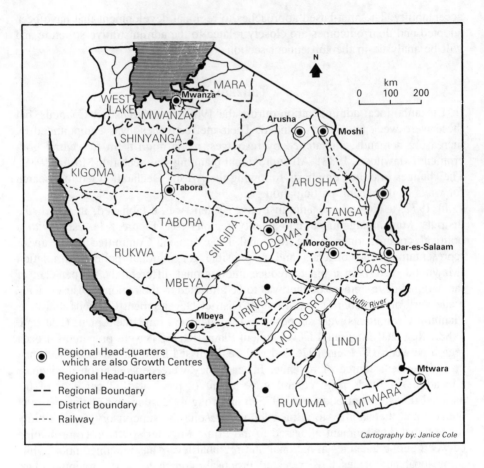

Figure 8.1 Tanzania: administrative boundaries and growth centres

some leverage over decision-making in order to ensure its own reproduction (Samoff, 1979).

A similar conclusion is reached by Asmerom (1984) with respect to Village Councils that, he concludes, are dominated by party and government officials, reduced in effectiveness by the lack of trained personnel and manipulated by well-to-do members of the village population. The latter interpretation is only partly borne out by an analysis of three villages in the Dodoma Region, where effectiveness of the state apparatus was related not just to control by the relatively wealthy but also to the outcome of struggles for influence and power by contending factions within the villages (Thiele, 1986).

In 1980 it was decided that decentralization to a regional level had not worked and, in 1982, the District Councils were reconstituted with a greater proportion of elected representatives than had been the case in the Regional Councils (Mawhood, 1983). Although the Regional Development Committees have been retained, they have only a co-ordinating and consultative role. Because of Tanzania's economic difficulties, particularly the loss of revenue from export crop production, increases in the price of imports and requirements for debt repayment, development budgets have become increasingly reliant on foreign funds (from 25 per cent in 1974–5 to 64 per cent in 1988–9) (Mtui, 1988, p. 15). Since 1982, funds have been distributed to district level, either directly or via the regions, in accordance with criteria and procedures that vary from year to year. It is difficult to see how districts will find the expertise to prepare plans or the funds to finance them. Their autonomy is very limited (Kauzeni, 1988), and implementation continues to suffer from constant changes in the institutional framework and policy directions, and from the absence of a satisfactory co-ordinating framework at either regional or district level (Mtui, 1988).

Zambia

From independence to the early 1980s, in parallel with its attempts to plan and implement rural development projects, the Zambian government was see-sawing between centralization in order to maintain central control over resources and a balance between ethnic and regional claims, and decentralization in order to improve the implementation of development projects and policies by mobilizing local resources. Rural local and provincial administration was traditionally weak because of its lack of staff and financial resources. Attempts to strengthen provincial administration (see Figure 8.2) included the appointment of provincial ministers and the establishment of Provincial Planning Units. However, little progress has been made with respect to the avowed intentions of allocating a greater share of national budgets to the provinces. Government capital expenditure was stagnant in real terms throughout the 1980s. Within this, the share allocated to provinces fell from 16 per cent in 1980 to 2 per cent in 1987, and the meagre sums available from central-government revenue were far exceeded by funds from donor agencies (Sikabanze, 1988).

In 1981 an attempt was made to strengthen local administration by establishing Rural District Councils, with a mixture of appointed and elected members. The councils have wide responsibilities but have been given neither central resources nor the ability to raise local revenues in order to enable them to fulfil these (Chikulo, 1989). Assessments of provincial- and district-level administration are lacking and those that do exist rarely consider the form and functions of the local state explicitly (Rakodi, 1988). Analyses of the national state have not been applied to subnational political and developmental processes, despite their overgeneralized treatment of rural class structures and/or the role of the state. Assertions such as that made by Good (1986a) – that the state apparatus functions as a resource for exploitation by the bureaucratic bourgeoisie and is a means of redistributing income to this office-holding sectional interest – need to be tested.

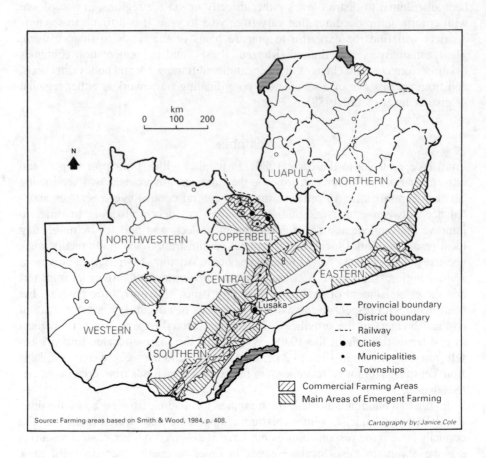

Figure 8.2 Farming areas and the administrative structure of Zambia in the 1970s

Zimbabwe

The pre-1980 subnational administrative structure consisted of eight provinces, with limited autonomy and responsibilities, and a fragmented local-government system. Relatively powerful urban and rural councils in the European areas of the country contrasted with numerous small African Councils in the communal areas. National development plans have referred to the need for regional autonomy and development to redress inequalities and achieve balanced growth.[1] The extent of co-operation at provincial level has increased, but while the African councils have gradually been merged to form District Councils (Figure 8.3) the intended merger with relatively wealthy Rural Councils which administer commercial farming areas has not yet taken place. The Provincial Councils and Administration Act 1984 attempted to establish a firmer basis for regional administration by establishing a hierarchical political and planning machinery from the national to the ward level. District and Provincial Development Committees are made up of both civil servants, representing central government, and local councillors. They are responsible for producing annual and five-year development plans intended to be a synthesis of central and local proposals. Provincial planning is expected to cut across sectoral policy sectors, preparing integrated development strategies and using these as the basis for submissions for expenditure under the Public Sector Investment Programme (Mlalazi, 1988; Rambanapasi, 1989).

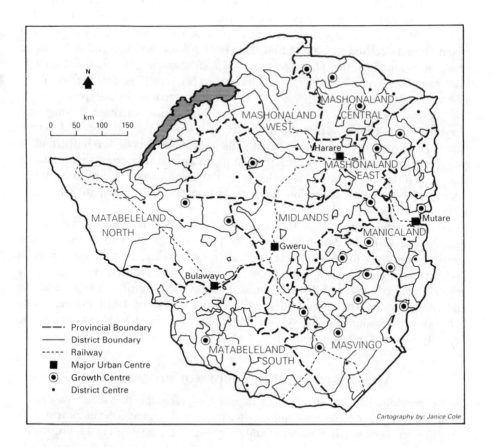

Figure 8.3 Zimbabwe: administrative boundaries and settlement policy

There has, however, been little support for strengthening provincial administration from central ministries, which have protected their sectoral independence and have not always seen eye to eye over areas of responsibility or approaches to planning. Notable protagonists are the Ministry of Finance, Economic Planning and Development (responsible for national economic development planning) and the Ministry of Local Government and Rural and Urban Development, to which power would be yielded if effective decentralization were to be achieved (De Valk, 1986). In practice, the provincial level of government has not been given autonomous revenue-raising capacity (Wekwete and Munzwa, 1986; Mlalazi, 1988). The failure to co-ordinate at central level may be alleviated by the recent establishment of a National Planning Agency to formulate proposals with regard to the spatial patterns of development investment (Mlalazi, 1988).

District Councils are responsible for the provision of health and education services, water supply, agricultural services and marketing. They also organize people for self-help projects, especially the reconstruction and construction of schools and health care centres. Women make a significant contribution to this in terms of labour, despite their relative exclusion from both formal and informal planning and decision-making (Chimedza, 1987). While attempts are likely to be made to strengthen the planning and financial capacity of District Councils (Helmsing and Wekwete, 1987), the relationship between these and the provinces is unclear. The involvement of a variety of donor agencies in different parts of the country and different programmes, with their project-oriented funding and varying approaches, increases the difficulty of co-ordinating budgets and expenditure (Wekwete and Munzwa, 1986). The urgent need both for redistributive policies at national level and greater local planning and implementation capacity is revealed by the spatial inequalities in the distribution of health services, commercial activities, extension services and accessibility found in Zinyama's (1987) district-level analysis.

In none of the countries has decentralization resulted in the devolution of power, mainly because of the concerns of party and government to maintain control and sectoral ministries to protect their areas of responsibility. In addition, financial problems and shortages of appropriate personnel have reduced the efficacy of regional or provincial and district administration. The evolution of representative and administrative structures at district level and below has tied local areas into a top down decision-making system rather than increased local autonomy and capacity for developmental activity.

Area-based rural development programmes

In the previous section, the limited capacity of regional, provincial or district administration to devise and implement integrated regional-development programmes has been noted. Parallel to and preceding these attempts at administrative decentralization, a variety of area-based approaches to rural and regional development have been adopted.

Tanzania

As none of the regional administrations established in 1972 had sufficient funds or expertise to formulate and implement Regional Integrated Development Plans (RIDEPs), donors were approached. Initially they were requested to prepare the plans, with a view to subsequently funding the projects identified and co-operating during implementation (Mosha, 1988; Ngasongwa, 1988).

By 1980, all the regions had RIDEPs, although only ten had received donor support for implementation. By 1988 donors were active in only three (Tanga, Iringa and Kilimanjaro – see Figure 8.1), having withdrawn from others partly because of the lack of progress and partly because of Tanzania's difficulty in reaching agreement with the IMF (Mosha, 1988; Ngasongwa, 1988). Kigoma Integrated Development Project,[2] for example, received eight years' funding under an IDA loan from the World Bank. The region is still considered to have economic potential for livestock and other development but, despite the funding, realization of this is constrained by poor infrastructure (*Daily News*, 16 August 1988). In a review of three of the integrated development projects (Iringa, Arusha and Kigoma), Kleemeier (1988) concludes that none was implemented successfully through the decentralized government structure because of the lack of autonomy at the regional level and scarcity of managerial resources. Nor were the projects able to engender local participation because villager willingness to contribute declined as the government demonstrated its inability to support such projects with staff and financial resources. Finally, none was able to adapt successfully to the changing and often adverse price structure.

The regional plans were supposed to be comprehensive reviews and strategy documents, with both spatial and sectoral dimensions, in which programmes and projects for the 1976–81 plan period would be identified (Mosha, 1988). In practice, the plans tended to analyse regions in isolation and varied in analytical quality, content and scope, partly because of the variety of donors involved and partly because the terms of reference were unclear. All tended to be one-off planning exercises. Some attempted comprehensiveness. Others drew together information as a basis for the later preparation of action plans, perhaps reflecting increasing donor awareness of the problems of comprehensive planning in situations of poor data availability and rapid change. Yet others were merely shopping lists of projects proposed by the districts, perhaps representing attempts at bottom up planning but failing to reconcile this aim with the need for integrated area planning. Although some have resulted in improvements to institutional and planning capacity and have contributed to regions' stocks of facilities and equipment, the implementation process has invariably been slow and incomplete.

Partly because of the neglect of spatial issues in the RIDEPs, zonal physical plans to be prepared by the Unit of Regional Planning in the Ministry of Lands, Housing and Urban Development were introduced in 1974. These were to be produced for five zones, each comprising two to three regions, and were to provide long term (20-year) guidance for integrated economic and physical development by means of physical development frameworks that will provide a basis for shorter term decision-making on the location of investment, planning of settlements and major infrastructure. Although not all the plans have yet been prepared, Mosha (*ibid.*) finds that few proposals have been implemented because the plans have no legal status; there have been some conflicts with existing proposals, as local bodies have not always been involved (Mtui, 1988); there is no administrative framework for implementation; and resources are lacking.

River basin planning was also introduced in 1975 with the establishment by Parliament of the Rufiji Basin Development Authority (see Figure 8.1), as a way of ensuring the integrated development of water and land resources of a single catchment area. Today, although four river basins have received some attention, development authorities, which fulfil mainly co-ordinating roles, have been established in only two (Mosha, 1988). These ineffective initiatives represent top down, centralized planning. Rather than merging physical concerns into the mainstream of economic and administrative decision-making, they were introduced as a technocratic approach in isolation from political and other administrative changes.

Zambia

During the Second National Development Plan period (1972–6) in Zambia, emphasis was placed on integrated rural development in Intensive Development Zones (IDZs). It was hoped that by concentrating resources in a limited number of areas with potential, self-sustaining growth would be achieved, and multiplier effects would benefit surrounding areas. Programmes of infrastructural provision, developing village centres, improved marketing arrangements and credit provision were implemented, initially in two areas in Eastern and Northwestern Provinces (see Figure 8.2). The lack of funds and skilled staff restricted the number of IDZs declared, and thus public support was limited (Sikabanze, 1988). Implementation was dependent on the availability of donor funds, giving rise to a variety of approaches. As a result, donor preferences, rather than domestic priorities, influenced policy and decision-making (Smith and Wood, 1984; Warren, 1988).

In 1979, IDZs were replaced by Integrated Rural Development Programmes (IRDPs), which were more widely spread in areas of both proven and undeveloped potential (Kalapula, 1988). While multilateral donors have supported agricultural programmes aimed at strengthening the agricultural research and extension capacity at the provincial level, IRDP activities, funded by bilateral donors (W. Germany, Sweden and the UK) have been directed at agricultural and rural development activities at the district level. By 1981, five such programmes had been established. However, only the Mpika IRDP had established a monitoring and evaluation system and so assessment of the socioeconomic impacts of the programmes has been almost impossible. Fostered by increases in credit and prices, as well as IRDP components, increased levels of production and sales were recorded in the Mpika District in the early 1980s. However, the need to substitute oxen for hoe technology to cultivate more than 4–5 ha imposed a ceiling on further growth (Warren, 1988). Although donors initially aimed to integrate their activities into the Zambian administrative system, the difficulty of doing so resulted in frustration and ineffectiveness. They gradually increased their organizational and procedural autonomy, which has helped the achievement of infrastructure provision objectives but has hindered replicability, sustainability and the development of local institutional capacity to prepare and implement development programmes (*ibid.*).

The impact of aid on a particular region is described for part of Northwestern Province by Crehan and von Oppen (1988). A project financed by an aid agency is shown not to be a predetermined package of measures but an arena of struggle in which various interests work to advance their cause by appropriating some component of the project. This may include building up political power by the judicious distribution of resources and funds (politicians), advancing careers (extension workers) or improving incomes or reducing insecurity (larger and smaller peasant farmers). Projects work, it is concluded, where the political needs for legitimacy by giver and taker coincide with each other and with the technocratic implementation capacity of the experts, and where the aims of the project are not contradicted by the general policy environment (Crehan and von Oppen, 1988; Elwert and Bierschenk, 1988).

Recognition of the dangers of proliferating IRDP management styles, different priorities and increasing autonomy led, in 1984, to requests for donors to work through government agencies, especially the District Councils. This was intended to strengthen the management and development-planning capacity of the latter, along the lines adopted by the UK-funded projects in Mpika District and Central Province. These are planned and implemented by existing local institutions (mainly District Councils), with the UK team acting as a catalyst. Following agreement on an annual programme, donor funds are made available locally. Expenditure has reflected district priorities and administrative capacity and has concentrated on roads, marketing, agriculture and water supply (rather than social facilities) to increase production, promote sustainability in local-revenue terms and reduce recurrent-expenditure implications. Although problems persist, the approach has increased the stock of physical infrastructure in the districts and increased their capacity for planning, implementation and monitoring. The second phase, which commenced in 1986, recognized that institution building is necessarily long term. Monitoring is undertaken by District Development Co-ordination Committees and the District Councils have drawn up proposals for local revenue generation (Mellors, 1987). A similar approach is adopted in the Zambian government guidelines for the formulation of District Development Programmes (Warren, 1988).

Zimbabwe

Before 1980, area-based planning in Zimbabwe consisted of a series of *ad hoc* approaches adopted by sectoral ministries. For example, in 1978, five Intensive Rural Development Areas (IRDAs) were identified by the colonial Ministry of Finance in communal areas either with the greatest population pressure or with unrealized agricultural potential (see Figure 8.4). Plans for these areas were produced by central government teams under the Agricultural and Rural Development Authority. In addition, in 1979–80 the Department of Physical Planning produced a traditional regional plan for Sebungwe Region, an area in the northwest straddling four districts, as it had power to do under the Regional, Town and

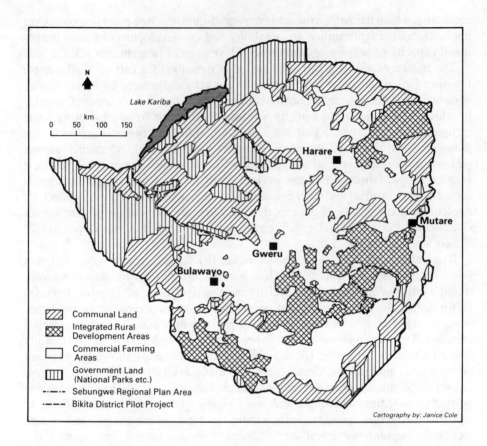

Figure 8.4 Zimbabwe: land classification and area development planning

Country Planning Act. However, the proposals were costly and overestimated the administrative capacity of government agencies so that, in the absence of an appropriate budgeting structure, little happened (Mlalazi, 1988; Rambanapasi, 1988).

The IRDA approach in general was characterized by a technocratic, top down orientation, reliance on external funding and end-state plans. Realization that this approach was inappropriate led to a gradual shift of emphasis to provincial and district planning and administration. Membership of the Provincial Heads Committees, previously comprised only of provincial officers of central government departments, was extended to include local authority officials, and planning committees at provincial and district levels were reorganized. Most of the provinces have now prepared development plans, although these are criticized by Rambanapasi (1989) as representing the product of a top down planning process aimed primarily at co-ordination of sectoral programmes. The United Nations Development Programme (UNDP) funded a pilot project in Bikita District in Masvingo Province in 1983 to develop a replicable district-planning and implementation system. Such a model has not emerged because of the extent of organizational and political change, the use of expatriate staff unfamiliar with local conditions and the difficulty of achieving co-ordination (Wekwete and Munzwa, 1986). Despite the emphasis on strengthening planning at the district and provincial levels, the idea of functional plans persists in the Zambezi Valley Master Plan, which incorporates the area covered by the earlier Sebungwe Plan (Figure 8.4).

Area-based approaches to development planning, whether based on functional planning aimed narrowly at increasing agricultural production or more broadly at achieving rural development objectives or on territorial planning associated with attempts at decentralization, have proved trickier than their proponents believed. Few have successfully embodied local participation. The building of institutional capacity, even for co-ordination, is clearly a long term process, given the ambivalent interests of central government agencies and the shortages of finance and resources.

National settlement policies

The final component of national regional policies to be discussed here is settlement policy, which has been strongly influenced by imported theories about the functions of urban centres in both industrial and agricultural development.

Tanzania

Tanzania's commitment to rural development has manifested itself both in attempts to reduce urban–rural income disparities and in a policy bias against Dar es Salaam. Policy with regard to incomes has worked in a variety of directions. That the net outcome has been to reduce urban–rural disparities is probably due to luck more than judgement. Thus,

> Since 1974–5 farm households' incomes have fared better than wages so that by 1982 farm households were as well off in real welfare terms as ... wage earners and far ahead of those minimum-wage earners whose households have to depend on their wages only. ... [Although] in absolute terms farm households did not improve their lot, subsistence production has made them more resilient to economic shocks than the wage earner.
>
> (ILO, 1988, p. 19)

The policy bias against Dar es Salaam is rooted in the perceived bad effects of political and administrative centralization, and manifest in attempts to control the city's growth (Armstrong, 1987), the decision in 1973 to relocate the capital, despite the cost and, since 1969, a policy of directing urban expenditure and industrial development to nine growth centres as counter-magnets to Dar es Salaam (Paddison, 1988) (Figure 8.1). It was hoped that economies of agglomeration and backward and forward linkages would develop and promote further growth, as predicted by standard growth pole theory.

In the 1970s, substantial investment only occurred in six of the nine growth centres (Obudho, 1983; Mosha, 1984), while Dar es Salaam continued to grow. The failure of most of the growth centres to attract industry and other economic activities has been attributed to their inadequate infrastructure, failure to allocate administrative responsibility, shortage of resources, lack of a clear and enforceable industrial-location policy and the inadequate criteria on which their selection was based (Mosha, 1988). The small overall volume of industrial investment, especially during the 1980s, as Tanzania's economic crisis and the international pressure on it to concentrate on agricultural production for export intensified, was also a significant explanatory factor.

In 1976, the centres of the regions were designated as growth centres, although these were later reduced to four industrial-growth zones, in response to the perceived over-ambitious aims of the earlier programme. Neither in 1969 nor in 1976 was it spelt out how industrial location and growth pole policies were to be implemented (Bryceson, 1984) and, while some manufacturing industries were successfully established in secondary urban centres, others were inappropriately located. Much investment in industry has been unsuccessful and unproductive and today all new public-sector industrial investment has been postponed, as only 30 per cent of existing industrial capacity is being used (Raikes, 1986; Skarstein, 1986). Despite the establishment of industrial estates in 16 regional centres by 1985, half of all employment in manufacturing was still concentrated in Dar es Salaam (Sawers, 1989).

There has, therefore, been little change in Tanzania's settlement pattern, although the proportion of the population living in urban areas more than doubled, from 5.5 to 11.5 per cent between 1967 and 1978, representing an average urban growth rate of 9.3 per cent per annum. The poor infrastructure and facilities available in Dodoma have deterred the intended relocation of many government departments and the primacy of Dar es Salaam, although reduced, continues (Obudho, 1983; Mosha, 1984). The city increased its share of the national population from 2.2 per cent in 1967 to 4.4 per cent in 1978 (O'Connor, 1988). However, the deterioration in urban living conditions during the 1980s has apparently resulted in a slowing down of its relative growth. Preliminary figures from the 1988 census reveal that the proportion of national population in the city had only increased slightly, to 4.8 per cent. The second urban settlement in 1967, Tanga, grew less rapidly during the 1970s because of the decline of its sisal-based regional economy. However, in more prosperous agricultural regions urban growth continued, in Arusha and later Mwanza (Kulaba, 1984; Sawers, 1989), encouraged by the availability of civil service jobs in regional centres following decentralization.

Zambia

Even in 1963, Zambia was the most urbanized country south of the Sahara apart from South Africa, with 21 per cent of its population in urban areas. This proportion had increased to 43 per cent in 1980. The most rapid urban growth occurred during the 1960s, accompanied by an increase in wage employment. In the 1970s, stagnation in the national economy and formal sector employment reduced the attractiveness of urban areas, especially the large towns, the economies of which were based on mining, industry and trade. Lusaka was the only partial exception. Nevertheless, urban growth continued at 6.7 per cent per annum between 1969 and 1980. By the late 1970s, Lusaka had ceased to be the fastest-growing urban area, while the four Copperbelt towns most dominated by mining grew at less than the national growth rate. Growth in other large centres continued despite the recession because of increased public sector employment, investment in state industries and new mine plant, and boundary extensions. Of the smaller towns, the growth of Kafue, the industrial new town established after independence, was most notable in the 1960s. However, during the 1970s, its rate of growth was exceeded by towns along the Tanzania–Zambia railway and some provincial capitals. Despite the more rapid growth of the smaller urban areas, Lusaka increased its dominance (22 per cent of total urban population in 1980, compared to 17 per cent in 1963) (Wood, 1982).

Sporadic attempts to decentralize industry to smaller urban centres have had only limited success and have often adversely affected the economic viability of the enterprises concerned. In the Third and Fourth National Development Plans industrial decentralization is set within a wider policy of rationalizing the settlement pattern as a basis for the efficient allocation of investment by classifying existing settlements into a hierarchy of growth poles/urban regions (Lusaka and the Copperbelt), ten provincial or regional centres of development, subcentres of development and district-level and village development centres (Kalapula, 1988) (Figure 8.2). Agencies have been established to foster rural and indigenous industrial development. It seems unlikely that such strategies will make much difference to patterns of investment or urban development.

Zimbabwe

Rural and regional development planning in Zimbabwe has also been concerned with the settlement pattern, specifically with trying to develop a hierarchy of urban centres under the generic name of growth points. The proportion of population living in urban areas, which may be under-enumerated, increased between 1969 (20 per cent) and 1982 (26 per cent), representing an overall growth rate of 5.4 per cent per annum and probably reflecting relatively slow growth up to 1977 and faster growth thereafter (Mutizwa-Mangiza, 1986; Simon, 1986). However, the settlement distribution changed little. Harare/Chitungwiza and Bulawayo maintained their dominance both in terms of population and industry (Helmsing, 1986; Riddell, 1988).

The 14 main centres are all situated in the commercial farming area, while urban settlements in the communal areas are inadequately developed, since pre-independence policy focused on decentralizing industry from Harare and Bulawayo to other main centres and confined urban development in the communal areas to low-level business/commercial centres. After independence, the programme for developing agriculture, reforming local government and providing infrastructure and services in the communal areas included ideas about the potential role of towns in fostering rural development. The aim was to develop a network of growth points and service centres based on theoretical assumptions about the integrating and developmental functions of a 'complete' urban hierarchy and the rural–urban linkages channelled through small urban centres. These assumptions had also underlain pre-independence policy recommendations (Hawkins Associates, 1980).

Both before and since 1980, the thinking has demonstrated conceptual confusion with respect to the assumptions of growth pole and central place theories (Wekwete, 1989a).

A network of 55 district centres and 450 rural service centres was identified, as part of a seven-tier national settlement hierarchy with towns and cities above, business centres and villages below. Thus rural service centres were intended to serve a population of up to 10,000, providing agricultural and social services, and acting as points for the distribution of agricultural commodities and inputs, in order to improve the production potential of their hinterlands and deter outmigration. District centres were to offer similar but higher-level services and ultimately to attract small scale industry, especially agroprocessing. Most of those selected are administrative centres, not necessarily those settlements with the greatest economic potential. Eighteen growth points were also defined. These were centres with a population of between 5,000 and 10,000, and 'an identifiable resource base capable of stimulating specific production and marketing activities and whose exploitation leads to rapid and sustained growth with development' (Zimbabwe Government, 1985, p. 13). Growth was to be based on processing local raw materials and eventually, it was hoped, these would grow into regional centres, diverting industry and migrants from the cities.

Funds were allocated for three phases of infrastructure provision between 1983 and 1986, 62 per cent for district service centres and 38 per cent for growth points. Infrastructure, especially electricity, water and telecommunications, has been installed, while trade and commerce have increased to some extent. Parastatal and private sector investment has been concentrated in centres serving productive agricultural regions, often pre-independence growth points (Zimbabwe Government, 1985; Wekwete, 1989a). The programme was partially suspended in 1987 because of its resource demands and the difficulty of demonstrating immediate development as a result of the investment. There is, however, considerable administrative and technical capacity for infrastructural development and this has been further strengthened by the establishment in 1986 of the Urban Development Corporation. Infrastructural investment yields evidence of commitment to rural development and is a relatively cheap and easy legitimizing device, but it has been typified by a narrow town planning approach rather than by attention to economic development and rural–urban relations.

Most discussions of manufacturing at the national level have considered neither the spatial distribution of industry nor small scale manufacturing (Ndlela, 1986; Riddell, 1988). Although decentralization of industry is considered desirable, its feasibility has not been seriously analysed, the policy is little developed and, until 1985, no financial incentives were available (Stoneman and Cliffe, 1989). As long as existing towns possess under-utilized infrastructure and few new firms are being established, centralization is likely to continue. Even where plants are established in outlying regions, few local linkages are generated (Helmsing, 1986). Large scale industry is relatively well established in Zimbabwe: it has pre-empted markets and is the source of inputs for rural enterprises that, when they expand, become potential competitors. While the development of centres makes available improved infrastructure and the public sector employment in them generates a market, these are insufficient to stimulate rural industry. Thus in 1982, 43 per cent of total net manufacturing output comprised the urban-based mass manufacture of consumer goods, many of which reach the rural areas as remittances from migrants.

Commerce and services are already developed to some extent in the larger urban centres in rural areas. Of the estimated 20,000 black business people in 1980, 15,000 were in the rural areas, mainly in trade and services (Munslow, 1980). Many of these are returned migrants who have been able to establish their businesses with the aid of a loan from an urban wholesaler in return for a contract to obtain supplies, the latter often obtained in turn from large urban manufacturers. Periodic markets in the 450 rural service centres are little developed because of the former settler regime's preference for fixed sites where the issue of licences to business people could act as a control mechanism, but it may be that such markets would be a more appropriate approach than investment in permanent commercial buildings. Gasper's (1988) evaluation of the growth centre programme shows that the provision of services and development of manufacturing in the rural areas must be considered within a more thorough knowledge of the way in which retailing, marketing and manufacturing are currently organized. Such knowledge should include awareness of the constraints imposed by the structure of these sectors on the establishment of small scale enterprises in outlying areas, including the dominance of production and markets by large scale manufacturers; the need to compete with established traders; and government restrictions on informal sector enterprises.

Conclusion

Perceptions of rural and regional development problems and their causes have varied over time in the three countries discussed here, although common themes are evident such as the desire to redress disparities both between urban and rural areas and between rural regions, the desire to mobilize local resources in the interests of achieving development goals and the need to maintain political support.

Explanations for the difficulties experienced in achieving agricultural production goals, especially by Tanzania and Zambia, include adverse terms of trade for export crops, climate, the mismanagement that has typified state intervention in agricultural systems and the ambivalence with which both countries have regarded the prospect of a prosperous and politically influential peasantry. In practice, the strengthening of central state control has taken precedence over a political ideology that stresses the populist nature of the state. The state rather than private enterprise has been seen as the main agent of development and progress.

However, control by the state over prices and marketing failed to secure a surplus sufficient to maintain the bureaucratic bourgeoisie in power in the 1980s in Tanzania (Stein, 1985), leading to a rolling back of the state under IMF and World Bank pressure. Paradoxically, this reduction of the role of the state may strengthen the hand of peasant farmers, via the market, while reducing the possibilities for equitable inter- and intraregional development. In Zambia, the agricultural crises and mismanagement of the 1980s seem more likely to have impoverished and disempowered the poor peasantry. The renewed favour with which some elements of the government view large scale private and joint-venture agriculture may not, however, achieve either production goals or political legitimacy. The Zimbabwean government has avoided agricultural production crises but only at the cost of soft pedalling in its attempts to redistribute land and other agricultural resources away from the large scale commercial-farming sector to farmers in the communal areas: the use of restrictions imposed for ten years by the Lancaster House agreement as an excuse became invalid at the beginning of the 1990s, but the prospects for a more radical restructuring do not seem promising.

A variety of attempts have been made by governments to find an appropriate administrative and political system to implement central policies and generate local development efforts and to resolve the contradictions between centralization and decentralization of political and administrative power. The extent to which satisfactory solutions have been devised, in terms of administrative capacity and achievement of developmental goals, and the relationship of the systems devised and their outcomes to the political and economic class structure of the countries concerned, have been investigated in a number of research studies. However, like much neocolonial planning policy and practice worldwide, much of the material available is written as if planning were a technical activity, many of the solutions advocated are institutional and neither explanations nor solutions fully recognize the complex nature of the interests involved.

Integrated rural development programmes in limited areas, especially in Zambia and Zimbabwe, have not, as far as can be ascertained from the available evidence, realized the expectations of their planners, and have proved neither sustainable nor replicable. The building of semi-autonomous 'donor republics' in all three countries, while facilitating the achievement of short term objectives, has failed to develop the institutional capacity at district or provincial level to plan, implement or manage development activities. This failure is exacerbated by the continued dominance of sectoral policy and administration, central control over resources and decision-making and unwillingness to devolve powers to set priorities and generate revenue to subnational levels of government. In each case, the pre-occupation of the national state with consolidating its own power has been understandable, both in terms of the objective political and security situation each has faced and in terms of the constellation of interests represented in the state (Weitzer, 1984; Stein, 1985; Sithole, 1987).

Finally, regional development planning has focused on the national urban-settlement hierarchy and the perceived need for change in the distribution of cities, towns and villages, both to counter the supposed adverse effects of 'overconcentration' in the largest cities and to realize the potential developmental effects of fostering secondary and lower-order urban centres. These hypotheses about the role of settlements in regional development are invariably supported by rather slender evidence. In addition, with the exception of villagization in Tanzania, less progress has been made with implementation than hoped, although analyses of the assumptions on which the policies are based and reasons given for the lack of progress are sometimes rather superficial. In particular, the conception of a national urban settlement policy in essentially physical terms, with insufficient understanding of the economic and political forces shaping settlement distributions, has led to over-optimistic expectations of the scope for change.

Notes

1. Despite the emphasis of the national plans on regional development, neither of the recent collections of papers on economic, social and political development in Zimbabwe contains discussions of regional development policy and planning or of the system of local government (Mandaza, 1986; Stoneman, 1988).
2. Official Tanzanian government sources refer to Regional Integrated Development Plans. These were, in some cases, prepared as a framework for a set of projects. The latter were generally referred to by donor agencies as 'integrated projects', presumably to reflect common usage of the term 'project' to mean a manageable unit of activity with a specified budget. However, the terminology is often confused.

References

Armstrong, A. (1987) Urban control campaigns in the third world: the case of Tanzania (occasional paper no. 19) Department of Geography, University of Glasgow.

Asmerom, H. K. (1984) The Tanzanian Village Council: its present status as an agent of rural development, *Planning and Administration*, Vol. 11, no. 2, pp. 82–90.

Brebner, P. and Briggs, J. (1982) Rural settlement planning in Algeria and Tanzania: a comparative study, *Habitat International*, Vol. 6, no. 5/6, pp. 621–8.

Bryceson, D. (1984) Urbanization and agrarian development in Tanzania, with special reference to secondary cities (unpublished report) International Institute for Environment and Development, London.

Bryceson, D. (1988) Household, hoe and nation: development policies of the Nyerere era, in M. Hodd (ed.) *Tanzania After Nyerere*, Pinter, London, pp. 36–48.

Burdette, M. (1988) *Zambia. Between Two Worlds*, Westview/Avebury, Boulder Col. and London.

Chikulo, B. C. (1989) The Zambian Local Administration Act, 1980: problems of implementation, *Planning and Administration*, Vol. 16, no. 1, pp. 62–7.

Chimedza, R. (1987) Women and decision-making: the case of District Councils in Zimbabwe, in C. Qunta (ed.) *Women in Southern Africa*, Allison & Busby, London, pp. 135–45.

Cliffe, L. (1988a) The prospects for agricultural transformation in Zimbabwe, in C. Stoneman (ed.), op. cit., pp. 309–25.

Cliffe, L. (1988b) Zimbabwe's agricultural 'success' and food security in southern Africa, *Review of African Political Economy*, no. 43, pp. 4–25.

Colclough, C. (1988) Zambian adjustment strategy – with and without the IMF, *IDS Bulletin*, Vol. 19, no. 1, pp. 51–60.

Crehan, K. and von Oppen, A. (1988) Understandings of 'development': an arena of struggle. The story of a development project in Zambia, *Sociologia Ruralis*, Vol. XXVIII, no. 2/3, pp. 113–45.

De Valk, P. (1986) An analysis of planning policy in Zimbabwe (paper given to the Workshop on the Planning System in Zimbabwe), Department of Rural and Urban Planning, University of Zimbabwe, Harare.

Elwert, G. and Bierschenk, T. (1988) Development aid as an intervention in dynamic systems, *Sociologia Ruralis*, Vol. XXVIII, no. 2/3, pp. 99–112.

Freedman, D. H. (1985) Popular participation and administrative decentralisation in a basic needs-oriented planning framework: the case of the United Republic of Tanzania, in

F. Lisk (ed.) *Population Participation in Planning for Basic Needs*, Gower, Aldershot, pp. 127–47.

Gasper, D. (1988) Rural growth points and rural industries in Zimbabwe: ideologies and policies, *Development and Change*, Vol. 19, pp. 425–66.

Good, K. (1986a) Systemic agricultural mismanagement: the 1985 'bumper' agricultural harvest in Zambia, *Journal of Modern African Studies*, Vol. 24, no. 1, pp. 257–84.

Good, K. (1986b) The reproduction of weakness in the state and agriculture: Zambian experience, *African Affairs*, Vol. 85, no. 339, pp. 239–65.

Hawkins Associates (1980) *Rural Service Centres Development Study*, Whitsun Foundation, Harare.

Helmsing, A. H. J. (1986) Rural industries and growth points. Issues in an ongoing policy debate in Zimbabwe (occasional paper no. 2), Department of Rural and Urban Planning, University of Zimbabwe, Harare.

Helmsing, A. H. J. and Wekwete, K. (1987) Financing district councils. Local taxes and central allocations (occasional paper no. 9), Department of Rural and Urban Planning, University of Zimbabwe, Harare.

Hodd, M. (ed.) (1988) *Tanzania after Nyerere*, Pinter, London.

International Labour Office (1988) *Distributional Aspects of Stabilization Programmes in the United Republic of Tanzania, 1979–84*, Geneva.

Jackson, J. C. and Collier, P. (1988) Incomes, poverty and food security in the communal lands of Zimbabwe (occasional paper no. 11), Department of Rural and Urban Planning, University of Zimbabwe, Harare.

Jacobs, S. (1984) Women and land resettlement in Zimbabwe, *Review of African Political Economy*, no. 27/28, pp. 33–50.

Kalapula, E. S. (1988) Approaches to subnational planning in Zambia: changes in the planning environment consequent to internal and external factors (paper given at the Rural and Urban Planning in Southern and Eastern Africa (RUPSEA) Network Workshop on Reactivating the Role of Planning at Subnational Levels: Perspectives on Southern and East Africa, University of Zimbabwe, Harare, 13–15 September).

Kauzeni, A. S. (1988) Rural development alternatives and the role of local-level development strategy, *Regional Development Dialogue*, Vol. 9, no. 2, pp. 105–38.

Kleemeier, L. (1988) Integrated rural development in Tanzania, *Public Administration and Development*, Vol. 8, no. 1, pp. 61–74.

Kulaba, S. M. (1982) Rural settlement policies in Tanzania, *Habitat International*, Vol. 6, no. 1/2, pp. 15–29.

Kulaba, S. M. (1984) The role of small towns in rural development in Tanzania, in H. D. Kammeier and P. J. Swan (eds.) *Equity with Growth? Planning Perspectives for Small Towns in Developing Countries*, Asian Institute of Technology, Bangkok, pp. 571–85.

Kydd, J. (1986) Changes in Zambian agricultural policy since 1983: problems of liberalization and agrarianization, *Development Policy Review*, Vol. 4, no. 3, pp. 233–59.

Legum, C. (1988) The Nyerere years: a preliminary balance sheet, in M. Hodd (ed.), op. cit., pp. 3–11.

Makgetla, N. S. (1986) Theoretical and practical implications of I.M.F. conditionality in Zambia, *Journal of Modern African Studies*, Vol. 24, no. 3, pp. 395–422.

Mandaza, I. (ed.) (1986) *Zimbabwe: The Political Economy of Transition*, Codesria, Dakar.

Maro, P. S. and Mlay, W. F. I. (1979) Decentralization and the organization of space in Tanzania, in A. Southall (ed.) *Small Urban Centers in Rural Development in Africa*, University of Wisconsin, Madison, African Studies Program, pp. 274–85.

Mawhood, P. (1983) The search for participation in Tanzania, in P. Mawhood (ed.) *Local Government in the Third World: The Experience of Tropical Africa*, Wiley, Chichester, pp. 75–105.

McCall, M. and Skutsch, M. (1983) Strategies and contradictions in Tanzania's rural development: which path for the peasants?, in D. A. M. Lea and D. P. Chaudhri (eds.) *Rural Development and the State*, Methuen, London, pp. 241–72.

Mellors, D. R. (1987) Integrated Rural Development Programme: Serenje, Mpika, Chinsali – Zambia (paper given to the Sustainable Development Conference, International Institute for the Environment and Development, London, 28–30 April).

Meyns, P. (1984) The political economy of Zambia, in K. Woldring with C. Chibaye (eds.) *Beyond Political Independence: Zambia's Development Predicament in the 1980s*, Mouton, Berlin, pp. 7–22.

Mlalazi, A. (1988) The changing nature of subnational planning practice in Zimbabwe and its implications for planning education (paper given at the Rural and Urban Planning in Southern and Eastern Africa (RUPSEA) Network Workshop on Reactivating the Role of Planning at Subnational Levels: Perspectives on Southern and East Africa, University of Zimbabwe, Harare, 13–15 September).

Mosha, A. C. (1984) Towards decentralised urbanisation: Tanzania's experience, in H. D. Kammeier and P. J. Swan (eds.) *Equity with Growth? Planning Perspectives for Small Towns in Developing Countries*, Asian Institute of Technology, Bangkok, p. 586–92.

Mosha, A. C. (1988) A review of subnational planning experience in Tanzania (paper given at the Rural and Urban Planning in Southern and Eastern Africa (RUPSEA) Network Workshop on Reactivating the Role of Planning at Subnational Levels: Perspectives on Southern and East Africa, University of Zimbabwe, Harare, 13–15 September).

Moyo, S. (1986) The land question, in I. Mandaza (ed.), op. cit., pp. 165–202.

Mtui, M. U. (1988) Financing and implementation of subnational plans in Tanzania (paper given at the Rural and Urban Planning in Southern and Eastern Africa (RUPSEA) Network Workshop on Reactivating the Role of Planning at Subnational Levels: Perspectives on Southern and East Africa, University of Zimbabwe, Harare, 13–15 September).

Mumbengegwi, C. (1986) Continuity and change in agricultural policy, in I. Mandaza (ed.), op. cit., pp. 203–22.

Munslow, B. (1980) Zimbabwe's emerging African bourgeoisie, *Review of African Political Economy*, no. 19, pp. 63–9.

Mutizwa-Mangiza, N. (1986) Urban centres in Zimbabwe: inter-censal changes, 1962–1982, *Geography*, Vol. 71, no. 2, no. 311, pp. 148–51.

Ndlela, D. B. (1986) Problems of industrialisation: structure and policy issues, in I. Mandaza (ed.), op. cit., pp. 141–63.

Ngasongwa, J. (1988) Integrated rural development in Tanzania, *Manchester Papers on Development*, Vol. IV, no. 1, pp. 101–22.

Nilsson, P. (1986) Wood – the other energy crisis, in J. Boesen, K. J. Havnevik, J. Koponen and R. Odgaard (eds.) *Tanzania: Crisis and Struggle for Survival*, Scandinavian Institute for African Studies, Uppsala, pp. 159–72.

Obudho, R. A. (1983) National urban policy in East Africa, *Regional Development Dialogue*, Vol. 4, no. 2, pp. 87–110.

O'Connor, A. (1988) The rate of urbanisation in Tanzania in the 1970s, in M. Hodd (ed.) *Tanzania After Nyerere*, Pinter, London and New York, NY, pp. 136–42.

Paddison, R. (1988) Ideology and urban primacy in Tanzania (discussion paper no. 33), Centre for Urban and Regional Planning Research, University of Glasgow.

Raikes, P. (1986) Eating the carrot and wielding the stick: the agricultural sector in Tanzania, in J. Boesen, K. J. Havnevik, J. Koponen and R. Odgaard (eds.) *Tanzania: Crisis and Struggle for Survival*, Scandinavian Institute for African Studies, Uppsala, pp. 105–41.

Rakodi, C. (1988) The local state and urban local government in Zambia, *Public Administration and Development*, Vol. 8, no. 1, pp. 27–46.

Rambanapasi, C. O. (1988) Subnational planning practice in Zimbabwe: an analysis of determinants of change. The case of regional planning (paper given at the Rural and Urban Planning in Southern and Eastern Africa (RUPSEA) Network Workshop on Reactivating the Role of Planning at Subnational Levels: Perspectives on Southern and East Africa, University of Zimbabwe, Harare, 13–15 September).

Rambanapasi, C. O. (1989) Regional development policy and its impact on regional planning practice in Zimbabwe, *Planning Perspectives*, Vol. 3, no. 4, pp. 271–94.

Riddell, R. (1988) Industrialisation in sub-Saharan Africa. Country case study – Zimbabwe (working paper no. 25) Overseas Development Institute, London.

Samoff, J. (1979) The bureaucracy and the bourgeoisie: decentralization and class structure in Tanzania, *Comparative Studies in Society and History*, Vol. 21, no. 1, pp. 30–62.

Sawers, L. (1989) Urban primacy in Tanzania, *Economic Development and Cultural Change*, Vol. 37, no. 4, pp. 841–60.

Sikabanze, M. A. (1988) Finance of subnational development and its implications for public planning - the Zambian experience (paper given at the Rural and Urban Planning in Southern and Eastern Africa (RUPSEA) Network Workshop on Reactivating the Role of Planning at Subnational Levels: Perspectives on Southern and East Africa, University of Zimbabwe, Harare, 13–15 September).

Simon, D. (1986) Regional inequality, migration and development: the case of Zimbabwe, *Tijdschrift voor Economische en Sociale Geografie*, Vol. 77, no. 1, pp. 7–17.

Sithole, M. (1987) State power consolidation in Zimbabwe: party and ideological development, in E. J. Keller and D. Rothchild (eds.) *Afro-Marxist Regimes: Ideology and Public Policy*, Lynne Rienner, Boulder, pp. 85–106.

Skarstein, R. (1986) Growth and crisis in the manufacturing sector, in J. Boesen, K. J. Havnevik, J. Koponen and R. Odgaard (eds.) *Tanzania: Crisis and Struggle for Survival*, Scandinavian Institute for African Studies, Uppsala, pp. 79–104.

Smith, W. and Wood, A. (1984) Patterns of agricultural development and foreign aid to Zambia, *Development and Change*, Vol. 15, no. 3, pp. 405–34.

Stein, H. (1985) Theories of the state in Tanzania: a critical assessment, *Journal of Modern African Studies*, Vol. 23, no. 1, pp. 105–23.

Stoneman, C. (ed.) (1988) *Zimbabwe's Prospects: Issues of Race, Class, State and Capital in Southern Africa*, Macmillan, London.

Stoneman, C. and Cliffe, L. (1989) *Zimbabwe. Politics, Economics and Society*, Pinter, London.

Svendsen, K. E. (1986) The creation of macroeconomic imbalances and a structural crisis, in J. Boesen, K. J. Havnevik, J. Koponen and R. Odgaard (eds.) *Tanzania: Crisis and Struggle for Survival*, Scandinavian Institute for African Studies, Uppsala, pp. 59–78.

Therkildsen, O. (1986) State, donors and villagers in rural water management, in J. Boesen, K. J. Havnevik, J. Koponen and R. Odgaard (eds.) *Tanzania: Crisis and Struggle for Survival*, Scandinavian Institute for African Studies, Uppsala, pp. 293–317.

Thiele, G. (1986) The state and rural development in Tanzania: the village administration as a political field, *Journal of Development Studies*, Vol. 22, no. 3, pp. 540–56.

Warren, D. M. (1988) A comparative assessment of Zambian Integrated Rural Development Programs, *Manchester Papers on Development*, Vol. IV, no. 1, pp. 89–100.

Weiner, D. (1988) Land and agricultural development, in C. Stoneman (ed.), op. cit., pp. 63–89.

Weitzer, R. (1984) In search of regime security: Zimbabwe since independence, *Journal of Modern African Studies*, Vol. 22, no. 4, pp. 529–57.

Wekwete, K. H. (1989a) Growth centre policy in Zimbabwe, *Tijdschrift voor Economische en Sociale Geografie*, Vol. 80, no. 3, pp. 131–46.

Wekwete, K. H. (1989b) *Rural Land Resettlement Programme in Post-Independent Zimbabwe – A Preliminary Review*, Department of Rural and Urban Planning, University of Zimbabwe, Harare.

Wekwete, K. H. and Munzwa, K. M. (1986) The role of external development agencies in promoting district and provincial planning and development – a case study of Masvingo Province (paper presented to a Workshop on the Planning System in Zimbabwe, University of Zimbabwe, Harare, 4–7 February).

Wood, A. (1982) Population trends in Zambia: a review of the 1980 census, in A. M. Findlay (ed.) *Recent National Population Change*, Special Publication of the Institute of British Geographers Population Study Group, London, pp. 102–25.

Zimbabwe Government (1985) Planning and management of human settlements, with emphasis on small and intermediate towns and local growth points in Zimbabwe (speech by the Leader of the Zimbabwe Delegation to the Eighth Session of the United Nations Commission on Human Settlements, Kingston, 24 April–10 May).

Zinyama, L. M. (1987) Assessing spatial variations in social conditions in the African rural areas of Zimbabwe, *Tijdschift voor Economische en Sociale Geografie*, Vol. 78, no. 1, pp. 30–43.

9

THE DISTRICT FOCUS POLICY FOR RURAL DEVELOPMENT IN KENYA: THE DECENTRALIZATION OF PLANNING AND IMPLEMENTATION, 1983–9

Marcel M. E. M. Rutten

Introduction

In 1983 the government of Kenya introduced a new policy for planning and implementing rural development. This strategy, known as the District Focus Policy for Rural Development (DFP) transfers considerable responsibility from ministerial and provincial headquarters to the district-level officers. Responsibility for general policy and the planning of multi-district and national programmes remains within the ministries. In this way a complementary relationship between the top down sectoral approach at the central level and the integrated, horizontal bottom up approach at the district level is sought. This chapter tries to analyse the progress made so far with reference to the decentralization of planning and implementation in Kenya.

Historical background of development planning in Kenya

Planning in Kenya can be traced back to the mid-1940s 'when the British Colonial Government called upon the Heads of Departments and Provincial Commissioners "to prepare plans" for post-war recovery' (Vente, 1970, p. 26). Funds made available changed the government policy of maintenance and control by adding the dimension of *development*. Outside experts as well as newly established special committees contributed to the plans for development used within the decision-making process. Their 'plans' were mainly restricted to the British tradition of budgeting – What is going to be spent on what and when, and where will the finance for it come from? This concept of 'good housekeeping' dominated the British administrative system transplanted into the colonies. Although the system was based on indirect rule (making use of the prevailing indigenous administrative or authority units) planning was still mainly a task of the central authorities.

At independence in 1963, Kenya inherited this strongly centralized and vertically integrated development administration and planning machinery. Shortly after independence some attempts were made to change the administrative set-up and concept of planning. A Ministry for Economic Planning and Development was established and the First National Development Plan (1966–70) was produced. In 1965 the government released sessional paper no. 10, which stated that:

(1) Planning is a comprehensive exercise designed to find the best way in which the nation's limited resources – land, skilled manpower, capital and foreign exchange – can be used.
(2) Planning cannot be done effectively unless every important activity is accounted for and every important decision-maker involved.
(3) Planning will be extended to provinces, districts, and municipalities, so as to ensure that in each administrative unit progress towards development is made.

(Republic of Kenya, 1965, pp. 1, 49, 51)

Besides the notion of scarcity of resources, other factors also supported the higher interest in careful planning. It was politically necessary for the new leaders to treat development of the young independent state as a very important issue, while the hope of foreign aid favoured the production of plans. This was connected with an equivalent expectation on the side of the donors (e.g. the World Bank); indeed, the first plans produced were mainly the result of findings and recommendations of various World Bank missions to Kenya.

In 1967, the Special Rural Development Programme (SRDP) experiment started in six selected pilot areas. It was a first attempt towards a more horizontally oriented form of development planning and administration. However, project development was disappointing, implementation was slow and interministerial co-operation never reached levels permitting an integrated local approach. The project was phased out entirely by 1977. SRDP failed because of unclearly formulated objectives and terms of action, on the one hand, and problems of political nature at the local administrative level on the other. Unselective data collection also proved confusing rather than helpful (Rondinelli, 1982; also Sterkenburg, this volume). Nevertheless, 'SRDP identified the need for improved skill levels, increased staffing and infrastructure support and commitment by central ministries to the concept of planning at and for decentralized levels' (Cohen and Hook, 1987, p. 79). SRDP also stimulated changes in the government administration: the District Development Committee (DDC), a civil service body, and the post of District Development Officer (DDO) were introduced.

Although the need for a decentralized form of planning was recognized in the mid-1960s, the Second National Development Plan (1970–4) was still mainly a product of the central planners. The Third Plan (1974–8) attempted to revive the concept of district planning by the preparation of forty local plans, one for each district. 'Labelled "District Plans" they were in fact largely written at province and headquarter levels by Provincial Planning Officers and expatriate advisers' (*ibid.* p. 81). An evaluation of all forty plans concluded that they contained too many proposed projects, failed to set clear principles, lacked detail needed by operational ministries and failed to merge with the national budgetary system. These experiences were used to improve district-level planning during the period of the Fourth National Development Plan (1979–83).

First, the government identified constraints to be overcome, namely,

(1) insufficient numbers of trained DDOs;
(2) inability of many DDCs to manage the district planning process;
(3) inadequate Treasury guidelines on budget ceilings;
(4) skepticism by provincial and district planners over the use of district plans by central ministries;
(5) poor liaison between planners and operating ministries in developing sectoral recommendations;
(6) insufficient generation and sharing of data to carry out planning exercises;
(7) limited participation by popular representatives in the DDCs; and
(8) financial information systems at the ministerial level that made it difficult to disaggregate expenditures to the districts.

(Ibid. p. 82)

Second, with funds and advisers from abroad (USAID and Harvard Institute for International Development) attempts were made to address these constraints: more DDOs were trained, guidelines for the district plans were provided and planning and budgetary links between district and national allocation processes were established. This time the district plans were indeed written at district level and their content exceeded the former shopping list of project proposals. Still, a real integrated district plan could not be produced mainly because knowledge of how to integrate local- and national-level planning was still lacking and, furthermore, the DDC had no authority to require action or co-operation from the operating ministries.

The environment for planning changed dramatically by the end of the 1970s due to national and international factors. With a change of government the Office of the President started to play a more direct and interventionist role in the economic business of administration and development. The government also recognized that the open nature of Kenya's economic system made it extremely sensitive to fluctuations in international economic conditions. During 1976–7, there had been a massive rise in the world price of coffee, Kenya's leading export product. As a result, several development projects were started and the government greatly increased its participation in an extensive range of private companies. However, due to the second oil crisis of 1979 and a drop in export revenues (low price of coffee) the budget and balance of payment deficits both increased dramatically. Kenya had to turn to economic policy discussions with such international institutions as the IMF and World Bank. Structural adjustments were needed to counterbalance the negative developments. Kenya reacted to the IMF and World Bank advice with several sessional papers and development plans. A document of major importance was the 1982 report of the Working Party on

Government Expenditures, which drew attention to public sector inefficiencies in the parastatal organizations and to the Treasury's inability to exercise adequate financial control. The Working Party (*ibid.* pp. 52–3) also reviewed the planning machinery and stated that

> There is a lack of a sharp, carefully co-ordinated focus on rural development at district level. There is too much emphasis on the provision of services and too little emphasis on involving the people and their resources in the development process. Yet, because officers in the field identify more with their superiors in Nairobi than with the people of the district, even the provision of services is carried out negligently and without dedication to or respect for the people being served. Distance precludes the adequate enforcement of discipline and accountability. Family, farm and national development all suffer as a result.

The conclusions and recommendations of the working party later formed the foundation for the District Focus for Rural Development Policy.

The District Focus Policy for Rural Development

As has been seen, since independence the call for decentralized planning has been heard and some pilot programmes were undertaken. The efforts of the 1970s taught some valuable lessons and pointed out the major constraints to be solved. However, in practice primary emphasis on centralizing economic development efforts still dominated Kenya's development strategy during the 1970s.

This changed dramatically with the introduction of the District Focus Policy for Rural Development in July 1983. Backed by the Office of the President, this new approach was introduced as a major initiative in the process of decentralizing planning and implementation of development in every district of Kenya. This resulted in changes to the (local) administrative set-up and to the planning and implementation process.

Kenya's administrative structure

Administratively, Kenya is divided into seven provinces encompassing forty districts, excluding the extra-provincial area of Nairobi. The districts are further subdivided into divisions, locations and finally sublocations. The so-called Provincial Administration forms the backbone of this administrative system, through a hierarchy of Provincial and District Commissioners, District Officers (division), Chiefs (location) and Assistant Chiefs (sublocation). The Provincial Administration is the main executive arm of the central government, operating under the direct supervision and authority of the Office of the President. The other technical ministries (Agriculture, Livestock, Education, Health, Water) are usually represented by their own officers in the various territorial units. These representatives are directly responsible to their ministry headquarters in Nairobi. A district also contains a system of local administration, comprising local authorities in the form of county councils whose boundaries usually coincide with those of the districts, and frequently of one or two urban councils.

Under the District Focus Policy the DDC has been chosen as the foundation of the decentralized rural development. This body is composed of the following:

- District Commissioner.
- District Development Officer.
- Departmental Heads of all ministries represented in the district.
- Members of Parliament.
- District KANU executive officer (the only Kenyan political party).
- Chairmen of local authorities.
- Clerks of local authorities.
- Chairmen of Divisional Development Committees (DvDC).
- Representatives of development-related parastatals.
- Invited representatives of non-governmental organizations (NGO) and self-help groups.

The DDC will meet at least four times a year. Technical support for its activities, including preparation of plans and management and implementation of projects, is provided by the District Executive Committee (DEC), which meets at least once a month. Membership of the latter is restricted to the senior officers and representatives of local authorities and parastatals.

The office of the District Development Officer is being strengthened by the introduction of the post of Assistant District Development Officer and by the establishment of a District Planning Unit (DPU). The unit's objectives are to serve as a secretariat to the DEC for day-to-day co-ordination of planning and implementation work. Several other (technical) posts will be attached to the DPU. In total, the DPU will have sufficient personnel to assist the departments not only in planning and monitoring but also in such technical activities as costing of projects or technical appraisal of project proposals. Beside appointing these special development officers, redeployment of technical staff from the centre and the provinces in order to raise the number and quality of the officers attached to the various technical ministries at district level, is one of the objectives of the District Focus Policy.

Alongside the DDC and the ministries, every district has several special purpose subcommittees operative in a specific field of development. These include the District Agricultural Committee, District Education Board, District Land Control Board and District Joint Loans Board.

At divisional level the Divisional Development Committee (DvDC) is operative. The DvDC's composition is a copy of the groups present at the DDC. The Locational Development Committee (LDC) and Sublocational Development Committee (SLDC) represent the grassroots-level community and are responsible for discussion of community needs and the initial identification of projects and activities to address those needs.

Planning and implementation under the District Focus Policy

In the past, planning and implementation in the district were seriously hampered by the dichotomy between the Provincial Administration and the technical ministries. The former had responsibility for the actual implementation of local programmes and projects but had no direct control over either the staff involved or their budgets, while the latter had power but no responsibility for overall policy, only for their particular specialized programmes.

The District Focus Policy specifically attempts to solve this main structural problem of the vertical integration of the ministries and the horizontal integration desired in the district. 'Responsibility for the operational aspects of district-specific rural development projects has been delegated to the districts. Responsibility for general policy and the planning of multi-district and national programmes remains with the ministries' (Office of the President, 1987, p. 1).[1]

District planning and co-ordination are carried out through a process that includes regular meetings of the development committees at each level of the district administrative structure and through the preparation of two basic planning documents: the five-year District Development Plan and the annual annex to that plan. Project suggestions made by the (sub)local communities are forwarded to the DvDC for selection and setting of priorities, the DDC making the final selection for the district as a whole. These project proposals are then included in the five-year District Development Plan and in the annual annexes to the plan. The objectives of the annex are

1. to show the progress made in implementing the District Development Plan;
2. if necessary, to adjust the plan with reference to available budgets or new projects; and
3. to convey district development priorities to the operating ministries for incorporation in the Forward Budget and Annual Estimates.

The DFP designers thought that effective and efficient implementation of the proposed projects would be enhanced by increasing responsibility at district level with respect to financial management, supplies management and by training of the officers on general principles of management and administration as well as on other district-focus related subjects.

Financial management

As a result of the District Focus Policy, the ministerial budgets are nowadays disaggregated on a district-by-district basis so that each DDC can anticipate the resources available for district-specific activities; ministries are obliged to base their programmes and budgets on the district's own plans and priorities; and guarantees should be given as to the funding of district-specific budget ceilings. Figure 9.1 shows the Annual Budget cycle and the several steps involved between identification and implementation of projects.

'The district focus strategy does not necessarily expand available resources beyond current levels. Funding of district activities remains under operating ministry control' (Cohen and Hook, 1987, p. 84). However, by transferring Authorities to Incur Expenditures (AIE) at the beginning of the financial year to the district level, the strategy tries to overcome unnecessary delays in implementation as in the past when AIEs were transferred in an *ad hoc* way from Nairobi via the provincial departmental heads towards the districts. Also, the District Cash Float Fund has been increased from K £1.3 million in 1983–4 to K £5.15 million in

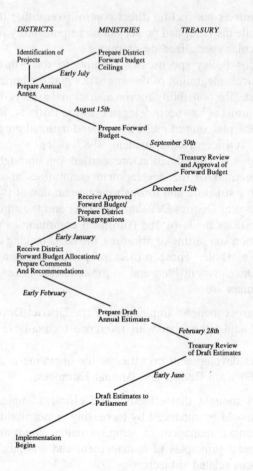

Figure 9.1 Kenya: Annual Budget cycle (*Source*: Office of the President, 1987)

1986–7 (1 K £ = 1 US $). This allows for more rapid payment of invoices.

One of the institutional changes with respect to financial management under district focus is the strengthening of the District Treasuries. Nowadays, accounting services are centralized under the District Treasury; the District Heads of Departments (AIE Holders) are supported by the District Accountants in preparing and examining vouchers. Internal audit work has also been improved.

Under DFP, the district is also responsible for planning and/or co-ordination of projects funded by foreign donors, local authorities (county or urban councils), self-help organizations and NGOs. Districts must identify financing for proposed projects, ensure that the funds have been committed and co-ordinate timely provision of the funds. Funds are limited and districts must set priorities for their use.

Supplies management

District Supplies Officers have been established in each district to handle the supplies and procurement operations of the district. District Tender Boards have been given more authority to process procurement for district-development activities. Previously, limits were set very low and most important purchases had to be handled centrally. Nowadays, inputs for district-specific projects, such as timber, sand, casual labour and the like should be procured locally wherever possible. District-produced inputs are allowed a 5 per cent cost bias.

District Focus training

A special National Training Strategy for District Focus has been prepared to ensure that all personnel involved in the implementation of District Focus, including the development committees down to the grassroots level, receive the required training. All District Commissioners have followed a full-time three-month course on management and administrative issues outside the country. Heads of Departments are trained under the responsibility of five major training institutions in Kenya. The District Heads of Departments themselves should train the Divisional Officers and the latter the Locational Officers by organizing seminars and workshops at those levels.

The District Focus Policy in Kajiado District

Evaluating the effect of the policy of District Focus for Rural Development is not an easy task. This is due partly to the DFP-induced changes in administration and planning after 1983 (e.g. District Annexes) that make comparisons over time difficult. Moreover, it is not always clear if new developments can be traced back to the incentives of the DFP or any other process. Nevertheless, the remainder of this chapter attempts to provide at least an overall impression of the effects of the DFP in the fields in which it has been operative since 1983. Data from Kajiado District, located in southeast Kenya (see Figure 9.2), and overall statements made by officers involved in the DFP have been used in this study.

Kajiado District (22,000 km^2) is divided into 4 divisions, 20 locations and 54 sublocations. These numbers are not fixed, and the status and number of administrative units are adjusted almost every year.[2] It is a semi-arid area inhabited by approximately 240,000 people who are mainly involved in nomadic pastoralism and rain-fed and irrigated agriculture.

Planning and implementation of projects in Kajiado District

Before the introduction of DFP, project implementation rates in Kajiado District were often very low. Reasons given are various, including a lack of participation by local people, poor project identification and a shortage of manpower and funds (especially in relation to the long shopping lists presented by the DDC) (see also Wisner, 1988).

Project-implementation rates under DFP are no longer available on a five-year planning-period basis. However, some general remarks can be made with reference

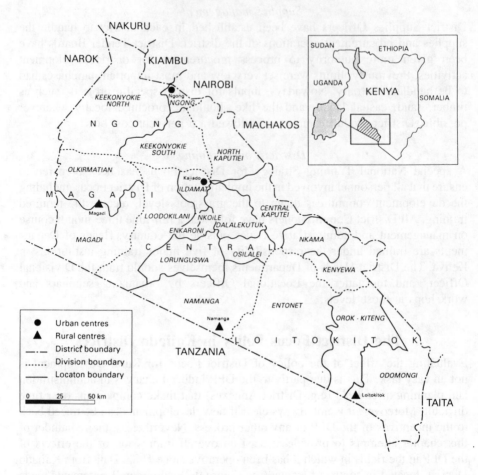

Figure 9.2 Kajiado District: administrative structure

to the presence or absence of obstacles within the planning and implementation process as far as Kajiado District is concerned. By comparing the district proposed projects for the financial year 1988–9 with the 1988–9 Printed Development Estimates (that is, the same list of projects after revision by the technical ministries and the Treasury in Nairobi) and finally the Kajiado District Development Plan Annex 1988–9 (showing what has actually been allocated in 1988–9) we are able to obtain an idea of the real change towards transferring responsibility under District Focus as far as planning and implementation are concerned.

So, in every financial year there are three moments at which the kind of projects and money allocated to them can be reviewed. These include the district proposals, the Printed Development Estimates as presented to the Kenyan Parliament and the Final Budget Allocation. Tables 9.1, 9.2 and 9.3 show a comparison of these moments.

Table 9.1 Comparison of District Proposed Forward Budget (DPFB), Printed Development Estimates (PDE) and Final Budget Allocation (FBA) of the Kajiado District Development Budget 1988–9 (in K £)

Sector	DPFB (K £)	(n.)	PDE (K £)	(n.)	FBA (K £)	(n.)
Administration	240,000	4	—	0	—	0
Home affairs	86,750	3	10,000	1	10,000	1
Finance	20,000	3	137,000	3	137,000	3
Agriculture	551,502	6	—	0	97,126	3
Health	2,250,000	20	460,000	4	145,000	4
Public works	18,563,946	12	—	0	3,568,150	6
Tourism/wildlife	10,000	1	—	0	—	0
Livestock	171,632	22	—	0	20,176	3
Cultural/social services	288,129	5	80,000	1	72,500	2
Information/broadcasting	17,000	1	—	0	—	0
Water	388,000	15	265,500	11	220,000	12
Environment	209,890	16	59,500	4	—	0
Co-operatives	120,000	2	—	0	2,120	1
Commerce	131,000	3	—	0	—	0
Energy	500,000	2	500,000	2	500,000	2
Education	320,000	3	20,000	1	20,000	1
Total	23,867,849	118	1,532,000	27	4,792,072	38

Note
(n.) = number of projects.
(*Source:* Author's own derivations based on MNDP, 1987, 1988a; ROK, 1988.)

In the Printed Development Estimates, projects are also mentioned that were not proposed by the district. They have been added to the list of proposals by the technical ministries or the Treasury in Nairobi. Table 9.2 shows these added proposals as well as the number and amount of money involved in the final allocation of these new proposals. Also in the Final Budget Allocation one can find proposals that were not present in either the District Proposed Forward Budget or the Printed Development Estimates (Table 9.3).

Figure 9.3 is a graphical reproduction of Tables 9.1, 9.2 and 9.3. It shows three different phases of the planning process that, since the introduction of District Focus, should in principle give more or less the same picture as far as funds and type and number of projects are concerned. However, from the tables it can be concluded that District Focus in Kajiado District has so far not been able to streamline this process in a satisfactory way. The budget requested by the DDC (approximately K £24 million) to finance 118 projects is rewarded with the provision of K £6 million to fund 82 projects.[3] However, only 38 of these projects were originally proposed by the Kajiado District Development Committee.

Table 9.2 Comparison of Printed Development Estimates (showing projects *not* proposed by the district) and the Final Budget Allocation of the Kajiado District Development Budget 1988–9 (in K £)

Sector	DPFB (K £)	(n.)	PDE (K £)	(n.)	FBA (K £)	(n.)
Planning/national development	—	—	77,500	5	92,500	5
Agriculture	—	—	1,000	1	—	0
Health	—	—	290,867	5	280,867	5
Public works	—	—	1,153,750	2	—	0
Livestock	—	—	60,000	1	81,500	1
Water	—	—	13,248,000	22	276,500	16
Environment	—	—	200,508	2	—	0
Co-operative	—	—	12,500	2	21,500	2
Industry	—	—	12,000	1	—	0
Total	—	—	15,056,125	41	752,867	29

Note
(n.) = number of projects.

(*Source:* Author's own derivations based on MNDP, 1987, 1988a; ROK, 1988.)

Table 9.3 Final Budget Allocations of projects *not* proposed by the district or presented in the Printed Development Estimates of the Kajiado District Development Budget 1988–9 (in K £)

Sector	DPFB (K £)	(n.)	PDE (K £)	(n.)	FBA (K £)	(n.)
Agriculture	—	—	—	—	42,500	1
Health	—	—	—	—	25,000	1
Livestock	—	—	—	—	278,000	1
Cultural/social service	—	—	—	—	26,500	4
Water	—	—	—	—	253,500	3
Environment	—	—	—	—	10,500	1
Education	—	—	—	—	106,750	4
Total	—	—	—	—	742,750	15

Note
(n.) = number of projects.

(*Source:* Author's own derivations based on MNDP, 1987, 1988a; ROK, 1988.)

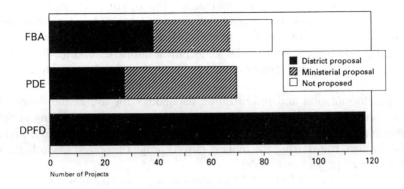

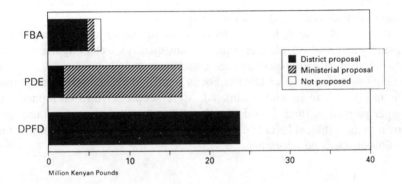

Figure 9.3 Kajiado District: 1988–9 Development Budget

Moreover, the actual allocation to specific projects was also very often not in accordance with the district's wishes.

Several, sometimes very specific, reasons are responsible for this discrepancy between the proposals made by the district and the final allocation by the technical ministries and the Treasury in Nairobi. It is thought that the lack of budget ceilings, in particular, led to a huge demand (some 10 per cent of the National Development Budget) by Kajiado District (representing only 1 per cent of Kenya's population) for the financial year 1988–9. On the other hand, it seems that the number of projects included by the technical ministries in the Printed Development Estimates and Final Budget Allocation is sometimes even higher than the number of district proposals (see, for example, the water sector). Foreign donors or international financing institutions (e.g. the World Bank), in particular, are responsible for the allocation of funds *not* present in either the District Forward Budget Proposal or Printed Development Estimates.

Information for analysing the planning procedure from (sub)locational to divisional and district level is not available. However, several Kajiado District officers stated, with reference to the District Focus Policy in general, that

- identification of projects is still problematic. The idea of involving the grassroots level (sublocation, location, division) does not work, due to a lack of guidelines and specific capabilities at these lower levels. Also, feedback within this project identification process between DDC, technical ministries and the lower administrative levels is poor. The production of long shopping lists of projects is still common;
- project formulation and prioritization handbooks for use at the district level have been developed (e.g. Bethke, 1983). However, priority-setting at district level comprises mainly a process of going along the lists of the four divisions and, in turn, transferring the highest-ranked project proposal towards the final DDC list of project proposals;
- co-ordination by the DDC of NGO projects does not take place. Contacts with NGOs are mainly restricted to occasions in which these organizations seek for assistance (e.g. in acquiring land for constructing a building);
- local contractors are not always available. If they are available their quality and/or charges are sometimes not acceptable;
- transferring of officers to the districts is still problematic as far as accountants (quantity and quality), planners (quantity and quality), water engineers (quantity) and supplies officers (quantity) are concerned; and
- training of officers under District Focus is still going on and the magnitude of this task seems to be underestimated. More people have to be trained and for longer periods of time. It is doubtful whether the approach of training trainers down to the sublocal level is a sound strategy. External inputs from the training institutions will be necessary.

Conclusion

Inclusion of the regional levels in the 'planning' machinery of the Kenyan government dates back to the beginning of the 1960s. However, in those days the Provincial and District Commissioners acted mainly as an implementing force of the plans and decisions made in Nairobi. In the course of time, the administrative structure at district level was changed by the introduction of several civil service bodies, such as the development committees at district, divisional and (sub)locational level. Also, recognition grew that deployment of regional 'planning' to solve the implementation problem *after* all decisions had been made centrally offered less chance of success than regional planning proper, which also includes the regional representatives in the whole planning process. Moreover, national planning normally gives emphasis to homogeneity and tends to ignore diversity in different physical, geographical and economic regions of the country. Being closer to the field, the officers at the (sub)district level are generally more in touch with the problems and potentialities of the regions and local people than are officials at headquarters. Although this was recognized for some time, it seems that the economic crises of the early 1980s, which hit Kenya's economy very severely due to higher oil and lower coffee prices, provided the ultimate stimulus to implementation of a decentralized form of planning. Besides measures of macro-economic policy, the government focused attention on public sector management. One of the conclusions was that the planning machinery was badly performing and had to be improved. In future, especial attention had to be given to the district level.

So in 1983, with very strong political support, a District Focus Policy was introduced that led to several structural modifications:

1. Supervision and disciplinary authority of the District Commissioner over the district ministerial staff increased, mostly by virtue of the enhanced authority of the District Treasury, the District Tender Board and the newly established District Executive Committee, all of which fall directly under the responsibility of the District Commissioner.
2. Under District Focus, senior officers are transferred from Nairobi to the districts and more junior civil servants are recruited to expand the manpower base within the districts. In addition, the Kenyan government started a huge training programme to uplift the quality of the (planning and technical) officers at all district levels. This is an ongoing process.

Looking at the achievements so far, it must be stated that considerable progress has indeed been made. An improved district planning system has been established. Guidelines for financial and supplies management have been developed. The quality of the District Development Plans and the efficiency of public services are said to have increased. There is an increased awareness that local-level decisions are important and that an integrated approach at district level is a far more viable approach than the old top down system of planning and implementation. It seems that there can be no return to the former process of planning and implementation, but is real fulfilment of the objective of effective decentralized and participative planning and implementation any nearer?

Information from Kajiado District suggests that the way ahead is still long. It is clear that several prerequisites for the efficient functioning of the District Focus Policy have thus far only taken shape on paper and not in practice. Some of the constraints mentioned by the Fourth National Development Plan (1979–83) before introduction of the District Focus Policy still exist. The major differences with respect to the kind and number of development projects proposed by Kajiado District and the funds ultimately approved and allocated to these projects by the technical ministries are too great. The role played by these ministries is sometimes very vague. Overall sectoral policy guidelines are very often lacking. Without active and proper support of these government bodies, District Focus will end up being still-born. In particular, the lack of information on budget ceilings hampers the real transfer of responsibility towards the districts. But even if the budget ceilings are properly taken care of by the technical ministries, the flexibility of planning and implementation at district level is still limited. For example, a district such as Kajiado, which is mainly oriented towards livestock and agriculture, has traditionally had a very low budget for industrial projects. Initiatives by the district in this direction will be hampered as a result.

A possible solution to improve flexibility of planning and implementation could be the introduction of a single unified District Budget, covering all the various ministry programmes in the district, and administered by a District Manager with a large measure of independent power and authority over all field officers. Some experience in this respect is being gathered nowadays through the Arid and Semi Arid Lands programme (ASAL), which has been operative in Kajiado District since the middle of 1987. This programme is an integrated rural development programme funded by the Dutch government. It is closely linked to the planning and implementation machinery of Kajiado District. It provides the DDC and the technical ministries with an overall additional budget allocation of 10 per cent. Moreover, this money can be allocated in a flexible way. Planning, budgeting and implementation are all in the hands of the district officers.

The dependency on foreign donors as far as the Development Budget of Kenya is concerned sometimes also interferes with proper district planning as it creates unforeseen cash flows and (the possible start of) new projects within the district. The opposite (withdrawal of funds) also happens and is of a more serious, disturbing nature.[4] The fact that district focus training workshops and seminars at divisional level are funded by various foreign donors on a more or less *ad hoc* basis is considered to be an undesirable situation.

Effective local participation is limited and will remain so for many years to come because of constraints in the administrative structure and political system, as well as the semi-nomadic way of life and the high level of illiteracy of many people in the district.[5] Participation in the planning process is mainly restricted to bureaucratic participation by officers and councillors at the (sub)locational and divisional levels. Some regard the role of politicians to be crucial in bringing about indirect participation of the people. For example, Delp (1981, p. 120) states that 'the local politicians should play a role of watch-dog in planning by seeing that fairness prevails and that civil servants follow-up in district discussions. What is needed is the appropriate forum, and the access to information embodied in the district level planning process.' In contrast, Tostensen and Scott (1987, p. 112) state 'it is difficult to see how in principle a political system, based in the final analysis on patronage and rapidly shifting factional alliances, can provide an entirely adequate foundation for a decentralized system of authority which presupposes rational decision-making, grounded on universalistic principles of administrative management'.

Under DFP, the role of the local political leaders and (sub)chief is not very clearly formulated. They seem to use this unclarity to dominate and politicize the project-identification process. That this is not a new phenomenon is also stressed by the Working Party on Government Expenditures (1982, p. 37): 'political considerations have at times affected the selection of projects and their regional location in the past and may do so in the future. These factors have also influenced the location of schools and health centres within districts and at times have resulted in uneconomic projects.' This seems to have created a general distrust in the civil service of political intrusions into the planning process: 'often a politician is viewed as an adversary whose main interest is government investment in his/her constituency, to improve the prospects of re-election' (Makokha, 1985, p. 37).

Participation of popular representatives in the DDC (especially compared to other countries in East Africa) is very poor since membership comprises approximately 85 per cent civil servants, 10 per cent representatives of parastatals and only 5 per cent politicians. In the author's opinion, this situation forces the politicians to 'go underground' and use their informal connections in the district or at a central level to achieve their own specific objectives.

Indeed, for Kajiado District it seems that the role of the politicians does not end at the district boundary. All in all, this possible disadvantage could probably be dealt with if real participation in planning and implementation by the local people were to occur. Adult education, awareness-raising, government assistance in implementing self-help projects and the like are the key issues to be involved besides a larger number of popular representatives at the DDC. Prospects of achieving that seem remote at present.

Some problems of District Focus will not be very difficult to overcome, while others of a more structural, political-financial nature could, in the end, lead to the frustration of this process of decentralization, as happened with the Special Rural Development Programme in the 1970s. In an economic sense, as long as resources are centrally controlled, the districts are in fact not accorded any significant autonomy. Moreover, the District Commissioner, whose central role in policy implementation is reinforced, is ultimately answerable to the Office of the President. The DFP might therefore be better described as a decentralization of financial control, as recommended by the working party, rather than a devolution of power from the centre. It is hoped that the outcomes of District Focus Policy will be valued positively by local officers and local people in the years to come. If so, increasing local participation as well as the introduction of a District Budget could and should be the next important ingredients for a real transfer of power towards the district.

Notes

1. Projects that are primarily intended to serve one district are referred to as 'district-specific' and are to be identified, selected, planned and implemented at the district level. Examples of district-specific projects include a village water system, a rural access road, a youth polytechnic and the like.

2. In April 1989, the DDC decided to establish a fifth division (Mashuru) comprising some (former) locations of Central Division.

3. Due to some very big projects, the total amount of money allocated for projects in Kajiado District for the year 1988–9 is even higher than it would be on average. One big water project (costing K £12 million) blurs the picture of the DPE. Most of these major projects are funded with foreign money (as grants or loans).

4. Approximately 50 per cent (30 per cent grants and 20 per cent loans) of Kenya's Development Budget over the years has been derived from foreign support.

5. It is thought that districts suffering from high levels of illiteracy and lack of communication will encounter more difficulties in raising levels of effective participation. For example, Kajiado District has a 12 per cent literacy level among the adult population. Approximately 10 per cent of the people are living in small towns.

References

Bethke, K. W. (1983) *Small Projects for Rural Development: Selection and Formulation Guidelines*, Ministry of Finance and Planning, Rural Planning Division, Government Printer, Nairobi.

Cohen, J. M. and Hook, R. M. (1987) Decentralized planning in Kenya, *Public Adminis-tration and Development*, Vol. 7, pp. 77–93.

Delp, P. (1981) District planning in Kenya, in T. Killick (ed.) *Papers on the Kenyan Economy: Performance, Problems and Policies*, Heinemann Educational Books, Nairobi, pp. 117–27.

Hyden, G. (1983) *No Short Cuts to Progress: African Development Management in Perspective*, Heinemann/University of California Press, London/Berkeley.

Jreisat, J. E. (1988) Administrative reform in developing countries: a comparative perspective, *Public Administration and Development*, Vol. 8, pp. 85–97.

Makokha, J. (1985) *The District Focus: Conceptual and Management Problems*, Africa Press Research Bureau, Nairobi.

Mbithi, P. H. (1982) *Rural Sociology and Rural Development: Its Application in Kenya*, Kenya Literature Bureau, Nairobi.

Ministry of Planning and National Development (1983) *Kajiado District Development Plan 1984–1988*, Republic of Kenya, Government Printer, Nairobi.

Ministry of Planning and National Development (1987) *Annex 1987/88 to the Kajiado District Development Plan 1984–1988*, Republic of Kenya, Government Printer, Nairobi.

Ministry of Planning and National Development (1988a) *Annex 1988/89 to the Kajiado District Development Plan 1984–88*, Republic of Kenya, Government Printer, Nairobi.

Ministry of Planning and National Development (1988b) *Kajiado District Development Plan 1989–1993*, Republic of Kenya, Government Printer, Nairobi.

Office of the President (1987) *District Focus for Rural Development* (revised March, 1987), Republic of Kenya, Government Printer, Nairobi.

Office of the President (1988) *District Focus for Rural Development – District Focus Training in the Districts 1987–1988*, Mutuga Development Institute, Kenya.

Republic of Kenya (1965) African socialism and its application to planning in Kenya (sessional paper no. 10), Republic of Kenya, Government Printer, Nairobi.

Republic of Kenya (1983) *Development Plan 1984–1988*, Republic of Kenya, Government Printer, Nairobi.

Republic of Kenya (1988) *Printed Development Estimates 1988/1989*, Vol. I and II, Republic of Kenya, Government Printer, Nairobi.

Rondinelli, D. A. (1982) The dilemma of development administration: complexity and uncertainty in control-oriented bureaucracies, *World Politics*, Vol. 35, no. 1, pp. 43–72.

Tostensen, A. and Scott, J. G. (eds.) (1987) *Kenya Country Study and Norwegian Aid Review*, Department of Social Science and Development/Michelsen Institute, Fantoft, Norway.

Vente, R. E. (1970) *Planning Processes: The East African Case* (IFO Afrika Studien 52) Weltforum Verlag, München.

Wisner, B. (1988) *Power and Need in Africa: Basic Human Needs and Development Policies*, Earthscan, London.

Working Party on Government Expenditures (1982) *Report and Recommendations of the Working Party Appointed by His Excellency the President* (Chairman: Philip Ndegwa), Republic of Kenya, Government Printer, Nairobi.

10

SYSTEM, STRUCTURE AND PARTICIPATORY DEVELOPMENT PLANNING IN INDONESIA

Allert van den Ham and Ton van Naerssen

Introduction

Indonesia in southeast Asia is one of the largest countries in the world. Its numerous islands, with a total land area of 2 million km^2, stretch 5,110 km along the equator. The island of Java is the heart of the country. It comprises only 6.6 per cent of the land area but contains more than 60 per cent of Indonesia's 175 million people, giving it a population density of around 800 per km^2. Java and the smaller islands of Madura and Bali constitute what is often called 'Inner Indonesia', in contrast to 'Outer Indonesia', the other relatively sparsely populated islands surrounding them. The population density of Kalimantan (27 per cent of the area of the country), for example, is only 12 per km^2 and of Irian Jaya, in the very eastern part of the country, even less (Figure 10.1).

Two main ecosystems are usually distinguished: wet rice cultivation, dominant in 'Inner Indonesia', and dryland agriculture, including shifting cultivation, in 'Outer Indonesia' (Geertz, 1963; Missen, 1972). These ecosystems to a large degree explain the differences in population density. About a quarter of the population could be considered as urban. The urban system is characterized by primacy. The position of the capital city, Jakarta, with 8 million inhabitants in its metropolitan region, Jabotabek, is outstanding. Other important cities, such as Bandung, Surabaya, Semarang, Medan and Palembang, have populations of 1–2 million each.

About half of the population live from agriculture, which contributes only 25 per cent to the GNP. Rice is the main staple and is cultivated on small parcels of land. The average farm occupies only 0.7 ha, while on Java the majority of

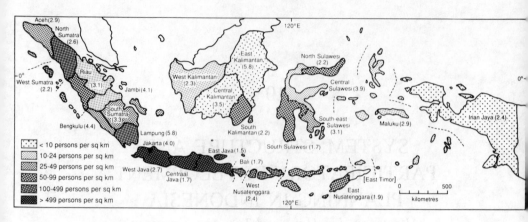

Figure 10.1 Population densities by province in Indonesia, 1980

the peasants cultivate less than 0.5 ha. Since colonial times, Indonesia has exported agricultural commodities like rubber, palm oil, coffee and tea, mostly originating from the outer islands. After 1965, however, the economy became highly dependent on two other sources: oil and foreign aid. Due to favourable prices on the world oil and gas markets, the country experienced a boom during the years 1973–85. At the beginning of the 1980s, oil and gas (a sector dominated by foreign investments) generated three quarters of the export earnings and one fifth of GNP. The flows of foreign aid are co-ordinated by the International Governmental Group on Indonesia (IGGI), consisting of the main members of the OECD and international donor agencies. In 1989 the IGGI donated and lent around US $4 billion. With the collapse of world oil prices and a debt ratio that increased to 40 per cent in the mid-1980s, the country cancelled many large development projects. It again focused its attention on the agrarian export sector and started export-oriented industrialization as well. In 1988–9, the industrial sector overtook oil and gas as the main source of foreign-currency earnings. The annual economic growth rate recently exceeded 5 per cent again. The GNP per capita is currently around US $550, one of the lowest in the region. There are large differences in the distribution of income, both by group and region. Poverty is particularly rampant in the countryside of 'Inner Indonesia'.

After independence in 1947, the country went through a rather confusing period under its flamboyant president, Sukarno. The period ended with a military coup in 1965, which claimed several hundred thousand victims. Since then, under the so-called 'New Order', Indonesia has become known as a politically stable country.

When reviewing the various options for regional development planning in Indonesia, decentralization would probably feature in almost all planners' recommendations. This is not surprising, of course, for how could a highly centralized, top down approach be successfully applied in a country as large and diverse as Indonesia? The physical and economic conditions outlined above seem to make a decentralized approach to development planning rational. But, as is shown later, this apparent rationality is surpassed by another rationality of a more political nature that aims to defend centralizing forces. In this chapter, the problem of regionalism in Indonesia and the centralized structure of government and development planning are first reviewed. Thereafter, attention is paid to efforts to decentralize planning and to implement a form of bottom up planning. Finally, the Strategic Development Framework (SDF) approach is presented, applied in the district of Sukabumi (West Java), as a concrete example of an attempt to plan from below. Although the approach is far from being a form of popular participation in planning, it is argued that it offers an opportunity for including popular consultation. Given the prevailing circumstances, it provides an alternative for concerned regional planners.

Bhinneka Tunggal Ika

It is clear that, in a country like Indonesia, the diversity in ethnicity, culture and religion is great. For centuries, coastal trade linked the islands to one another. Yet, strong unifying forces never evolved as occurred in India and China. The Netherlands East Indies, the entity created by the colonial regime, was an artificial one that, as a matter of fact, was achieved only at the beginning of the twentieth century when the Dutch succeeded in ending the Aceh War (1904). It is not surprising, therefore, that regional feelings in Indonesia are strong. On the other hand, separatist movements are not manifest nowadays, although they do exist for specific reasons in Irian Jaya and in East Timor.

The Dutch considered Irian Jaya as a separate entity adjacent to Indonesia, and it was only after intervention of the UN that it became part of the country in 1962. The original inhabitants are Papuans but the main economic sectors are presently in the hands of other ethnic groups. Moreover, transmigration (colonization) projects caused a large influx of Javanese settlers, which added to social tensions (Otten, 1986, pp. 155–84). East Timor was invaded in 1975. Since then it has been annexed as the 27th province of the country, but this has yet to be accepted internationally and a state of war still exists in the region. These two cases aside, one can say that the unified state of Indonesia is generally accepted. But a delicate balance still exists between central and regional forces, and the motto on the national coat of arms, 'Bhinneka Tunggal Ika', which means 'Unity in diversity', expresses this well.

Under colonial rule, Jakarta, or Batavia as it was called, developed as a focal point for governing the Netherlands East Indies. Although in some areas of the outer islands, especially in eastern Indonesia, the Dutch governed by indirect rule, the system was essentially a centralized one. During the war of independence, the Dutch favoured a federation of autonomous Indonesian states. By applying a divide-and-rule policy, they hoped to be better able to defend their interests in Indonesia. The forces of independent Indonesia, however, developed differently. As Ichalsul Amal and Nasikun (1988, p. 25) remark

Revolution and the war for independence in opposition to the existence of the colonial government was an extremely important process in the nation building of an independent Indonesia. This process created solid criteria for the type of national leadership, political attitudes, national values and objectives for the future, and the form of government that was ideal and appropriate for the Indonesian situation. All of this bore a single consequence, that is, that centralization, apart from all of its weaknesses and faults, was more dominant than decentralization in the spirit of Indonesian political life after Independence. Nation building, which appeared in the credo of a united nation, overcame the premises of decentralization, such as the principle of efficiency, rule of law, and, especially, community participation in government in the autonomous areas.

During the second half of the 1950s, the country faced several rebellions in Sumatra, Sulawesi and West Java, which had their roots not only in ethnic and cultural differences but also in discontent about Javanese domination of the political system and the distribution of resources. One of the grievances was that, while both mineral resources and agricultural commodities stem mainly from the outer islands, the revenues went to Java. None of the rebellions was successful, but they had considerable impact on the political constellation. In order to cope with decentralizing forces, the provinces were granted authority over all matters that were not legislated for at central-government levels, by way of the creation of provincial assemblies in 1957. They were also to be allowed to elect provincial governors. However, after only two years, Sukarno established a centralized structure more in line with his 'Guided Democracy', by Presidential Decree in 1959. More specifically, governors were to be appointed by the president.

In many respects, the *coup d'état* of 1965 represented a watershed in the history of independent Indonesia but the relationship between central government and the provinces retained its centralized character. As is shown in the next paragraph, the structure of interaction between the central and lower tiers contains some elements that reflect decentralizing tendencies. However, the actual workings of a structure depend on the wider sociopolitical system. Structuring government by legislation and defining the tasks and authorities of its constituent elements is one matter; implementing a policy is quite another, and relates to other dimensions of the decision-making realm. Vested interests, power relations and state ideology belong to the system that ultimately determines the value of a structure. At the national level this could be illustrated by the constitutional structure. Three parties are allowed to compete for the favours of the voters. The elected members of the People's Consultative Assembly, together with a sizeable number of appointed colleagues, in turn elect the president every five years. Yet, in practice, the role of the political parties in policy-making is limited compared to that of two other interlinked actors in the Indonesian political landscape: the military and the bureaucracy. Therefore it seems more appropriate to characterize Indonesia as a military bureaucratic state (Robison, 1978; MacDougall, 1982). While the structure suggests a form of parliamentary democracy, in reality it is used to legitimize the system that works the other way round. In fact, the core of the sociopolitical system consists of 'a nearly total control by the Presidency over both the army and the bureaucracy and over important sectors of the economy through state enterprises or trusted managers who are privileged with monopoly licences' (Schulte Nordholt, 1987, p. 53). The creation of new structures, whether meant to decentralize or not, will only make sense when accompanied by changes in the system. Keeping this relation between structure and system in mind, our attention is now turned to how regional development planning in Indonesia actually functions.

The structure of government and development planning

Four levels of local government can be distinguished: the province, the district (*kabupaten*), the subdistrict (*kecamatan*) and the village (*desa* or *kelurahan*). Large cities (*kotamadya*) are considered to be at the same level as districts. Jakarta, as the capital city, is an exception and possesses provincial status. During the process of the formulation of the first five-year plan, Repelita I (1969–74) (Repelita, meaning Rencana Pembangunan Lima Tahun), thought was again given to the relationship between central and local authorities and to the question of how to allocate development funds to the regions (Sanders, 1971).

The former was arranged by Law No. 5 of 1974, which defined the structure of local government. Provinces and districts are administratively autonomous regions, which means that they have elected representative bodies. The heads of the regions, respectively the governor and the *bupati* or mayor, are, however, appointed by the president. They make decisions and the influence of the regional parliaments is small. At the provincial level, the Provincial People's Representative Council advises, the governor decides and the president is the one to whom he is ultimately responsible. This arrangement places the governor in the ambiguous position of simultaneously representing central government, by virtue of his presidential appointment, and supposedly also the regional population, by virtue of his function. The same duality can be found at the level of implementation. In each province, central-government departments have branches (*kanwil – kantor wilayah* – meaning regional offices) that look after centrally administered and funded programmes. They report both to the governor and to Jakarta. Next to them, each province has its own sectoral agencies (*Dinas*) administering the regional funds. The same structures exist at the level of the *kabupaten* where the head of the district is responsible to the governor, and in the *kecamatan*. However, no representative council exists in the subdistricts. At the lowest level, the status of the *desa* is defined as semi-autonomous. The head of the village is chosen by the village and thereafter appointed as a government representative by the head of the district. A special committee, the Village Resilience Organization (LKMD, meaning Lembaga Ketahanan Masyarakat Desa) performs an advisory function. It usually consists of the local élite, village officials and other village representatives. Important matters are supposed to be discussed during meetings of the Village Community Council, in which all adults officially participate. All together there are 66,000 villages, 3,300 subdistricts, 300 districts and 27 provinces in Indonesia.

Development planning operates within this governmental structure. At the national level, BAPPENAS (Badan Perencanaan Pembangunan Nasional, meaning National Development Planning Board) in Jakarta is authorized to devise and to monitor the five year plans. It is here that co-ordination of sectoral planning from the many ministries takes place. In 1974, planning boards, called BAPPEDA (Badan Perencanaan Pembangunan Daerah, meaning Regional Development Planning Board), were set up at provincial level, followed by the creation of similar boards at the district level in 1980. These are known as BAPPEDA Tingkat II (level II, to distinguish from the boards at the first, provincial level). In other words, structures for horizontal, area planning took shape at the provincial and district level. In reality, however, the planning system is still vertical, with unequivocal dominance of top down sectoral planning from the ministries to the lower level governments. The above-mentioned representative branches in the regions are instrumental in this. This is where the conflict between structure and system becomes paramount. In theory, all development activities undertaken by the sectoral agencies should be co-ordinated and integrated by the governors and the district heads staff, via the BAPPEDAs. In reality, vertical links within the departments and ministries, defined mainly by the flow of funds and the career prospects of its staff, are proving to be much stronger than the horizontal relations with the regional authorities. The development strings in the regions are therefore pulled largely by those who control the purse: the central government departments and their representatives at lower levels, for their budget usually makes up the bulk (65–90 per cent) of all regional expenditures, leaving only a small portion for the regional policy makers.

Part of this small portion involves the direct allocation of the National Development Budget to the regions. During Repelita I (1969–74), procedures were established to transfer funds from the central to the lower levels. These subsidies were meant to increase the experience of regional governments in implementing development projects that should be their responsibility but for which they lack financial resources. They were also specifically meant to stimulate development in the rural areas. The allocation was arranged in a Presidential Instruction in 1970 and the programmes are therefore called Inpres (Instruksi Presiden) programmes. They comprise about one sixth to one seventh of the national development expenditures and involve two kinds of subsidies. The first is related largely to unspecified block grants to provinces, districts and villages (areal grants), while the second category, grants for specific sectoral projects in the regions (health, education), was added in the mid-1970s. It is clear that the opportunities for local decision-making in development planning are greater with the former than with the latter.

During the years of the oil boom, more development funds became available and Inpres programmes expanded steadily. But this growth was accompanied by a shift from areal to sectoral grants. While in the mid-1970s 85 per cent of the grants fell into the first category, ten years later the balance was 60:40 in favour of the latter (Booth, 1988, p. 11). So, in reality, Inpres moved more and more in the direction of sectoral projects, carried out by the regions but approved and paid for by the central government. When Inpres started, some observers considered it a promising initiative to give local governments more say in development planning at the lower levels. However, after two decades, one can only reach the conclusion that Inpres contributes but marginally to enhanced local decision-making.

Although possibilities for decentralized decision-making within the structure of government and administration do theoretically exist, in reality the planning system is strongly centralized and top down planning prevails. Governors, heads of districts and mayors have large powers, while the influence of provincial and district councils is limited. At the lowest level, the head of the village acts as a representative of the government and the village meetings are mostly a matter for the village élite.

There are several reasons why the system of decision-making doesn't allow for significant participatory planning. The historical background explains much. Feudalism and the accompanying attitudes had a long history, particularly in Java, while Dutch colonial rule strengthened hierarchical structures. Also, the diversity of ethnic groups, cultures and religions, often regionally vested, is great and at the central level there is always the fear of national disintegration. Moreover, for reasons of recent history, the present Indonesian government has successfully sought to maintain control of political, religious, labour and non-governmental organizations. Their role is severely curtailed, specially in the rural areas. 'In this way only the bureaucracy could serve as an instrument for communication between the government and the grassroots level' (Grijpstra, 1989, p. 3).

Nevertheless, there are some initiatives and efforts to delegate decision-making to the lower levels of the government and to accommodate pressures from the regions. In this respect, we should also mention the role of development co-operation. The World Bank and other multilateral agencies, together with the bilateral donors, have had a strong influence on development planning since the beginning of the New Order. These circles support initiatives to achieve decentralization and fund programmes initiating it. The main arguments are that when local authorities are involved, efficiency in planning and implementation of programmes and projects will increase. Also it is thought that the lower income groups can be reached more easily within a decentralized structure.

The Provincial Area Development Programme

In 1976, the Indonesian government introduced the Provincial Area Development Programme (PDP). It differed from the Inpres Programme because implementation was not a matter for the *Dinas* but part of the tasks of the BAPPEDAs. In 1977, USAID agreed to sponsor a number of PDP projects with US $32 million; later on other international donor agencies became involved in projects.

The USAID-sponsored programme was devised as an experiment for a period of ten years. The aims of the programme were twofold:

1. Training and institution building by developing expertise in local area and regional development planning at the provincial and district level. Government officials involved, especially the staff of the BAPPEDAs, should obtain experience by giving them responsibilities and funds, i.e. learning by doing.
2. Initiation of projects focusing on lower income groups in the rural areas, i.e. to increase the income of rural poor in the project areas, using the approach of integrated rural development.

In other words, the programme combined ideas of decentralization of planning with concepts such as target groups and basic needs. According to Ichalsul Amal and Nasikun (1988, p. 30), 'PDP was expected not only to be able to increase the ability of the regional governments to articulate regional needs, but also to be able to increase the ability of the entire regional development bureaucracy to develop a planning system from the bottom.' Hence, PDP was associated with bottom up planning approaches.

PDP was carried out in Jakarta, in 8 provinces and in 44 districts. The programme started in the provinces of Aceh and Central Java, was extended to Bengkulu, South Kalimantan, Nusa Tenggara Timur and East Java in 1979 and finally ended in 1980 with the provinces of West Java and Nusa Tenggara Barat. Technical assistance and training funds came from USAID, while the projects obtained joint funding from the government of Indonesia and USAID.

The strategy of the programme was to start with pilot projects that, if successful, would be duplicated elsewhere. The majority of the projects were located at the district and village levels. Using a combination of government statistics and surveys, the poorest districts within provinces were identified. Once these districts had been selected, determination of the poorest subdistricts followed. Within these subdistricts, the poorest villages were identified, after which the poor village families were selected for inclusion within the programme (Weinstock, 1989, p. 2). The several thousand small projects included many sectors, including food crops, estate crops, small scale irrigation, livestock, fisheries and small industries (Table 10.1). Usually, the projects were subsidized for the first year and if they continued, participants could borrow money in the successive years with the support of a credit system (MacAndrews, Fisher and Sibero, 1982, p. 99). The BAPPEDA staff and other government officials were involved in the selection, design, implementation and evaluation of the projects. So everything was in their hands from the start. Both long term and short term consultants were assigned to each of the 8 provinces to assist in project development and to provide training programmes in planning and management. These contained a strong practical component by linking theory with the experience gained in project implementation. So at the level of institution building, the capacities of the BAPPEDAs to supervise projects at the local and regional levels and to take responsibility in managing their own project's budget increased.

Did the PDP achieve its aims? Jim Schiller, who evaluated the programme for USAID, calculated the cost per beneficiary (including overhead costs) at Rp 250,000. This is low compared with other projects like the World Bank sponsored transmigration projects (Schiller, 1988, p. 52). It seems that the programme had a direct positive impact on the incomes of some 600,000 families, of whom between 56 and 88 per cent in each province belonged to the lowest 50 per cent income group. But this definition of low-income groups is, of course, fairly broad and one should like to know about the impact on the lowest-income groups (20 per cent). It would also be interesting to know about the effects on the 12 to 44 per cent in each province in the high 50 per cent income group.

Regarding the upgrading of BAPPEDAs and other government services, PDP is also considered to be successful. MacAndrews, Fisher and Sibero (1982,

Table 10.1 Sample pilot projects in four selected PDP provinces, 1979–80 and 1980–1*

| | Programme year | |
	1979–80	1980–1
Planning/evaluation	8	8
Training	4	12
Community development	6	9
Credit†	18	30
Rural industries (general)	5	7
Handicrafts	5	8
Appropriate technology	2	6
Irrigation	5	12
Agricultural integrated farming systems	7	9
Multicropping trials	3	6
Rice	3	8
Sugar cane	1	2
Coffee	2	2
Coconut	1	—
Rubber	1	4
Home gardens	2	4
Tapioca	1	—
Food storage	1	2
Fisheries	8	12
Agricultural resource stations	2	4
Animal husbandry	4	10
Total programme for all four provinces	89	155 ‡

Notes
* Bengkulu, South Kalimantan, Nusa Tenggara Timur and East Java.
† Credit covers projects in all areas (i.e. agriculture, home industries, animal husbandry, etc.).
‡ This includes some duplication, particularly in credit.

(*Source:* MacAndrews, Fisher and Sibero, 1982, p. 103.)

pp. 95–6) write that already 'by the early 1980s it was clear that it had achieved a significant degree of success and was becoming adopted as a model for local area planning in Indonesia'. If so, there is scope to start a substantially expanded programme in the near future. Although the regional development budgets after 1983 diminished due to cuts in the national development budget, they are still large compared to the PDP budgets. BAPPEDAs with experience could achieve a greater say in the allocation of these regional budgets but there are no indications that this will occur in the near future.

At present, a programme with some similar features to that of the PDP is being carried out in urban areas. The Integrated Urban Infrastructure Development Programme (IUIDP) started with Repelita IV (1984–9). It is based on the same idea of decentralization to obtain more efficiency in planning and to reach the poorer segments of the population. Such urban infrastructural works as roads, sewerage and public water are usually planned and financed by the Ministry of Public Works in Jakarta and urban authorities do not go beyond implementation. The IUIDP aims to change this situation and to delegate the planning and financing from the ministry to the municipalities. Under the IUIDP they are trained to formulate Urban Investment Plans in which infrastructural works are situated within a coherent structure and their financial implications are assessed. It is the intention that the municipalities will finance urban works from their own resources in future. The Ministry of Public Works will then be concerned mainly with giving technical advice. The IUIDP is financed by international agencies and implemented with assistance from foreign consultancy firms. IUIDP projects are currently being implemented in eight provinces of Sumatra, Sulawesi, Bali and Java. In the long term, decentralization will take place in about 500 towns and cities, spread over the whole of Indonesia. In the short term it would be interesting to see if this urban programme could be combined with planning in the surrounding districts and subdistricts.

Bottom up planning procedures

By Instruction no. 4 of the Minister of Home Affairs, an initiative was taken to establish a bottom up planning structure. Procedures were introduced to facilitate the inclusion in the budgets of proposals for development projects and programmes at village level. At this level, proposals are discussed by the Village Resilience Organization (LKMD) during annual village meetings. After approval by the Village Community Council, they are forwarded to the subdistrict. The intention is that the proposals are then thoroughly vetted in a joint meeting of the district BAPPEDA (Tingkat II) with sectoral offices in the subdistrict and all village heads. At this meeting BAPPEDA also announces which of the proposals from the previous year will be implemented in the current year.

In the next stage, proposals from the subdistricts are discussed by the BAPPEDA of the district and the sectoral agencies of this level. They can add their own proposals to the list, which needs to be approved by the District People's Representative Council. Thereafter, the proposals go up to the province, where BAPPEDA Tingkat I, together with the sectoral agencies of the province, screens and modifies the list. Finally, the National Development Planning Board in Jakarta reviews project proposals from all the provinces. In consultation and co-operation with the ministries concerned, it is decided which projects will be implemented in the following year.

Obviously, it was likely that many constraints would hamper proper decision-making along the intended lines. Not only did all the officials who had to participate in the process require training but also the sociopolitical climate mentioned earlier creates the basic difficulty that participation by the people is limited to the LKMD, which means in effect the '*tokoh-tokoh masyarakat*', or local élite. Furthermore, the administrative apparatus is not yet ready to activate the system countrywide. Actually, it seems that the procedures are being applied faster on the island of

Java than elsewhere. According to van den Ham and Hariri Hady (1988, p. 75), 'Each year hundreds of thousand people become involved in generating, processing and forwarding development proposals to higher levels. Functionally, however, most observers seem to acknowledge that in spite of the good intentions the centralised, sectoral and top down planning system has remained intact.' According to their observations, the majority of the proposals from the villages themselves are sectoral, concerning health care, education, religion, community welfare and the like. No integrated packages are forwarded, and the number of proposals referring to productive investments is usually negligible.

At the levels of the subdistrict and district one observes a shift from social to physical project proposals, probably due to the fact that the district officials try to formulate proposals in such a way as to suit the existing, top down sectoral policies. In other words, the heads of the subdistricts, and at higher levels the heads of the sectoral agencies, decide on which projects will be forwarded to the provincial administration, based on their assessment of which proposals have the best chances of attracting funding from the central government budgets. And, as a matter of fact, such projects as buildings and roads are easier to plan and implement than social projects that, for example, aim to increase the quality of education or community development! Again, at these higher levels, there exists a remarkable lack of proposals of a more economic nature. One of the reasons could be that even if such proposals are available, local governments are unable to think within frameworks of integrated area planning. Accustomed to top down procedures, it also appears that they find it difficult to accept proposals from the villages. The shift from social to physical projects means that, in practice, the villages receive things other than what they requested. It provides another illustration of the argument that simply creating structures for increased participation in decision-making does not guarantee effective implementation.

Seen from this perspective, it is of utmost importance to train the planners of BAPPEDA and the sectoral agencies concerned to think and to act in spatial terms. They should be able to analyse and integrate proposals from the lower level within an appropriate spatial development framework. If not, it is unavoidable that projects and programmes proposed to the central level will have a shopping list, sectoral and haphazard character. Moreover, planners should be open minded and responsive to the villagers' wishes and knowledge of their own living conditions and the potentials and constraints of their local environment. This applies especially in regard to low-income groups.

The Strategic Development Framework approach

An interesting recent initiative along these lines has been tested during the formulation of an integrated area plan for the district and municipality of Sukabumi in the province of West Java (Figure 10.2). The project opted for the Strategic Development Framework approach (SDF, meaning Kerangka Pembangunan Strategis, KPS). It was applied by the West Java Regional Development Planning Project 1986–8, a joint effort of West Java BAPPEDA, Dutch consultants and the Ministry of Home Affairs. The latter, by way of its Agency for Personnel

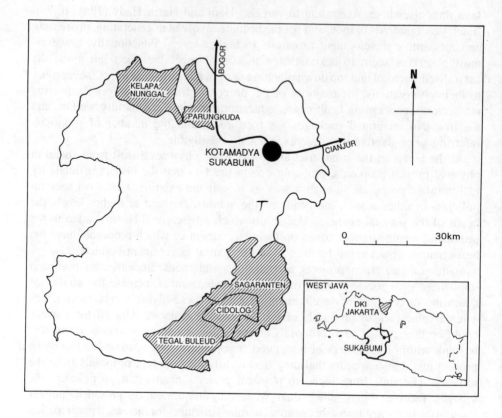

Figure 10.2 Sukabumi Development Region: Strategic Kecamatan

Education, was involved because of the training aspect.

The project aimed at strengthening the capabilities of the Regional Development Planning Office as well as at the identification of development priorities and the subsequent formulation of programmes and projects. A regional development plan for the years 1989–94 had to be formulated for one of the provincial development regions. Besides its inherent objective of presenting a consistent set of programmes and projects, this plan was also intended to serve as a prototype for future plans in other regions. Consequently, the methodology of plan formulation was seen as crucial. This underlines the significance of the concept of strategic development planning that has been used through the planning process.

The SDF has been born out of necessity. District authorities in Indonesia are usually not ready for their prescribed task of formulating a comprehensive regional development plan. Constraints are the lack of specific training in this field, limited development funds, lack of time horizon and shortage of regionally recorded data. The SDF approach therefore aims at deliberately limiting the working load of the planners to a rational selection of target areas, target groups and target sectors. This means that, for the time being, a number of topics are being left out of consideration.

An SDF covers a timespan of 3–5 years. It consists of a spatial framework wherein the needs and potentials as perceived by selected target groups are related to specific projects and programmes in some selected key sectors. This implies the selection of 'strategic' target groups, in terms both of deprivation and potential. Their perceived needs and wishes are investigated. Later on, backward areas and promising sectors are selected, omitting others. In this way, SDF tries to determine areas, groups and sectors that can provide an optimal contribution to the attainment of the two national development objectives of economic growth and social equity.

In the SDF approach, the target group, and within it the household, form the basis for planning. Once the needs and potentials as perceived by small farmers, landless agricultural labourers, fishermen, estate workers or small merchants are known, planners and target groups can together identify the development activities required to solve those problems or utilize the potential. The emphasis on the target group's perception of its own needs and potential as well as of the solutions and courses of action required, marks a new form of popular consultation at local level.

Of course, it does not comprise popular participation in the sense of taking part in, and being able to influence, the decision-making on the development of the community. However, this kind of policy formulation, based on direct consultation with the target group, is a major improvement over the present system in which planners single-handedly identify the development issues and matching policies. In addition to participation in the traditional annual 'bottom up' planning cycle, which in practice does not seem to benefit the weaker sections of the village population a great deal, the SDF therefore provides the target groups with another opportunity to exert influence on the direction of the development process in their immediate environment. By participating in the formulation of the SDF, they can join in defining the topics that are going to be tackled in the following years (van den Ham and Hariri Hady, 1988).

The case of Sukabumi: conditions and trends

At the request of the Indonesian authorities, the West Java regional team focused its attention on the Sukabumi Development Region. This region comprises the district and the municipality of Sukabumi (Figure 10.2). Located on West Java's southern coast, Sukabumi covers an area of 4,179 km^2 and contained a population of 1,827,000 in 1985. Since the construction of a detour of the 'Grote Postweg' in the nineteenth century, the mountainous area of Sukabumi has been firmly linked to the economy of West Java. Rubber and tea developed into its most important products.

Much has changed since the colonial days. When the estate sector more or less collapsed in the turbulent days after independence, Sukabumi was dragged along. Its decline did not stop after the New Order regime took over in the mid-1960s. Over the last decade, the economic growth of Sukabumi has been consistently slower than that of the province of West Java. In 1977 its contribution to the West Java GRDP was still 4.6 per cent; by 1986 it had dropped to 3.6 per cent. This compares unfavourably with its share of the population (5.9 per cent) and the land area (9.4 per cent). Per capita, the GRDP of Sukabumi in 1977 still amounted to 76 per cent of the provincial average (excluding oil and gas); in 1986 it had decreased to a mere 61 per cent. At that time, the per capita income in Sukabumi amounted to Rp 280,962, whereas an average inhabitant of West Java could count on Rp 457,659. These figures indicate that economic development has not been evenly spread across the province, and that a district such as Sukabumi could justifiably be considered a lagging development region within the province.

Development within Sukabumi has also not been spread evenly in space. Developments in the northern part have been strongly influenced by the presence of the Bogor–Sukabumi–Cianjur road, which links it to the national capital, Jakarta, only two hours' drive away. Moreover, this part possesses good soils and abundant water supplies. A string of centres developed in this area and, in time, it became the economic and administrative heart of the region. The southern part of Sukabumi, on the other hand, is characterized by less fertile soils and relatively limited groundwater availability. Nevertheless, the formerly sparsely populated south has increasingly been opened up for food-crop cultivation by landless farmers. The production capacity of the estates has deteriorated severely over the years.

Sukabumi is still largely a rural area. This is not only reflected by the small share of the population living in urban areas (21.5 per cent in 1985) but also by its economic structure. It is true that the share of the agricultural sector in the GRDP has decreased considerably from 44.6 per cent in 1975 but, at 35.1 per cent at present, it still ranks quite high. The dominance of the agricultural sector is still more apparent from the employment figures: in 1985 more than 377,000 people (56 per cent of the working population) derived their main income from agriculture. At the same time, these figures reflect the low value added per worker in the sector. Other important sectors are trade (21.2 per cent) and public administration (13.3 per cent).

The growth of agriculture has occurred mainly in the food-crop and smallholder estate-crop subsectors. To a considerable extent, this has been accomplished by extensification of the harvested area. Recently this trend has stopped abruptly, indicating that nearly all suitable land is now in use. In contrast, the planted area of the large private estates has declined by 6,884 ha, or 22 per cent between 1980 and 1986! This has partly been caused by the cumbersome procedures to renew land leases and obtain investment credits. But other, usually politically well connected private estate companies that did obtain land leases and credits, invested the funds in other, more profitable enterprises, leaving the estates idle. The population census of 1980 registered some 194,000 households as farming their own or someone else's land. Of these, only 22 per cent had access to an area larger than 0.5 ha, whereas 53 per cent had less than 0.25 ha. Another 148,500 households were classified as agricultural labourers.

The industrial sector in Sukabumi used to be dominated by small-scale enterprises. Since 1975 a transformation has taken place. Due to differential growth rates of large and medium industries that produce mainly for the provincial/national market, and the small ones that cater for the local inhabitants, the former dominate the output of the manufacturing sector. Yet small and home industries provide employment for 54,000 and the large and medium industries for only 9,500 workers. In spite of government intentions, the region's location (half an hour's drive from the Jagorawi toll road) and its endowment with favourable climatological conditions as well as great natural beauty, tourism is still a minor economic activity in Sukabumi. This is attributable to rather basic physical infrastructure that, while adequate to support the present level of activities, provides little stimulus to further economic development.

In environmental terms, Sukabumi has paid a high price for its economic development. Continued deforestation of the mountain slopes and delay of reforestation have resulted in erosion, loss of productive land, loss of water storage capacity in the soil, flash floods and a reduction of the river basin floods. Over-extended land use has aggravated these problems. Extensive quarrying has turned vast areas into unproductive wasteland. There is an increasing danger of land slides in zones of geological instability. Discharge of industrial waste is causing growing water and air pollution.

The above observations lead to a projection of a declining economic growth performance of Sukabumi; the annual rate of growth, which was still 6.4 per cent during 1982–6 and 5.9 per cent during 1985–9, is expected to slow down to only 5.4 per cent per annum during 1989–94. With economic growth in West Java at 7 per cent, the 1994 income per capita in Sukabumi will decrease further to only 57 per cent of that of West Java. With an average employment elasticity of 0.34, a 5.4 per cent annual growth in GDRP would create 69,000 new jobs during Repelita V, more than 21,000 short of the projected need. This means that only 76 per cent of all new entrants on to the labour market in Sukabumi can be absorbed.

With the present level of investment and routine expenditures, the quantitative aspects of primary education (pupils/class ratio, teachers/class) can be improved considerably. In lower secondary education, the picture looks bleaker. Without any additional investments, the government has the option of either increasing the number of pupils per class significantly and by doing so reducing the quality of education, or limiting the intake of new pupils, causing the enrolment to drop from 38.6 to 34.2 per cent. Conditions in higher secondary education are expected to remain at the same level. The budget allocated to vocational training is expected to remain small, thus hampering the development of a skilled workforce. No real attempt will be made to redress the relatively small participation of girls in secondary education. Further improvement of the health situation is very much a question of education, immunization, sanitation and other preventive measures. Due to financial constraints, the shortage of medical personnel will continue, as will the present emphasis on curative treatment. Finally, it is not expected that the present gap in development between the 'north' and the 'south' will be closed. On the contrary, the major construction activities on the Cirurug–Sukabumi–Cianjur through-road will only consolidate the pre-eminent position of the northern part of the development region.

It is clear that a substantial effort is needed to close the gap between the Sukabumi Development Region and West Java as a whole. As a matter of fact, it is not realistic to expect a substantial increase in both private and public investments. Maintaining Sukabumi's relative position to the province seems the maximum attainable goal for the coming period.

Would it be possible to reduce spatial and social imbalances within Sukabumi? In the 1989–94 Regional Development Plan, it is argued that the shortage of funds and investment also forms the main constraint to reducing the imbalances. Yet, by distributing the limited funds to the most pressing needs, it should at least be possible to prevent a widening of the gap as far as the accessibility to a number of public services is concerned. By carefully analysing the potentials and constraints of poor groups in backward subdistricts and distributing the funds accordingly, it should be possible to stimulate economic development in these areas, albeit on a very limited scale. Here the concept of strategic development planning can prove valuable.

Application of the SDF approach

The SDF for Sukabumi has been formulated by a combined team of planners from the BAPPEDA and the sectoral agencies. It has been observed that the sectoral agencies tend to be very reluctant to include the proposals into their budgets lest the BAPPEDA formulate a development plan on its own. By broadening the functional base of the team, the technical quality of the plan would be improved and the commitment of the sectoral agencies to implementation of the proposed projects and programmes be increased. To facilitate the formulation of the SDF by a local team of planners, a training-cum-planning programme has been designed (van den Ham, 1989a). Under the programme, the participant planners underwent a short period of classroom training, applied their recently acquired knowledge in the field, brought the data back to the classroom for further processing, received additional methodological guidance, did some more fieldwork and so forth. In total, there were five weeks of classroom training. In the end the participant planners produced a Strategic Development Framework that should be eligible for inclusion in the district Five Year Plan V (1989–94).

Based on the score of a number of indicators representing the potentials (soil types, land potentially to be opened up, increase in the industrial labour force, product diversification, savings, credit, etc.) and the equity conditions (*sawah/ person* ratio, quality and provision of roads, electricity supply, agricultural productivity levels, environmental conditions, service level, health, education, etc.), five target subdistricts could be identified. In addition, the two subdistricts in the Kotamadya have been selected.

Two of the rural subdistricts, Parungkuda and Kelapanunggal, are located in the north. Fieldwork in these reasonably accessible areas identified landless agricultural labourers (daily income Rp 9–1,200/day) and small farmers (<1 ha) as the main target groups. General problems observed are low health standards and a lack of infrastructure. The target groups are confronted by low production levels in irrigated rice cultivation as well as annual and perennial dry crops, and a lack of employment opportunities outside the agricultural sector. Potentials were identified for coffee, melinjo and hybrid coconut production, and inland fisheries as well as bamboo handicrafts. By contrast, the other three rural subdistricts, Sagaranten, Cidolog and Tegal Buleud, are located in the southeastern part of the *kabupaten.* The last, bordering the Indian Ocean, is relatively isolated. The problems and potentials of these areas are distinct from those in the northern areas. These predominantly agricultural subdistricts are less densely populated, the soils are generally of a poorer quality and, because of their more eccentric location, they have less potential to cater for the large market of Metropolition Jakarta (Jabotabek). In general, they suffer from a low level of health, education, water supply, electricity and other infrastructural services. More specifically, the target groups suffer from the low productivity of their agricultural enterprises, a less diversified production structure and poor educational background. Potentials for further development comprise the reuse of the extensive fallow private estates, improvements to the low productivity of smallholder estates and increases in rice, vegetables, fruit and livestock production. For the most southern subdistrict, Tegal Buleud, major potential for prawn hatcheries has been identified.

In general, the two subdistricts of the municipality of Sukabumi are characterized by a limited water supply and sanitation facilities, a lack of cheap housing and a slow industrial development. The SDF target groups in the *kotamadya*, small farmers, street vendors and labourers, experience low productivity due to pests and lack of irrigation, the lack of proper business facilities and readily available credit, and limited skills. By improving the modest irrigation system and providing essential inputs, the low agricultural productivity is expected to increase. Providing credit to small industries is expected to eliminate a major bottleneck for their expansion, whereas creating favourable business locations for hawkers would eliminate one of their main problems.

In the end, applying the SDF approach in Sukabumi provided a number of proposals tailored to the needs of selected target groups in a limited number of areas. This is not enough if the formulation of a regional development plan is at stake, for it does not take into account the needs of other disadvantaged groups living in somewhat more prosperous areas, or the potential contributions to regional development by sectors that are not the domain of the poor. But it does, however, provide a number of well-argued, concrete proposals to ameliorate the position of the poor in a coherent way. A future improvement would be to supplement the SDF with a more sectoral approach. Major potentials and problems outside the strategic areas would accordingly be identified, analysed and treated. The result would be a plan covering the main potentials of the region and the needs of its poor residents. Finally, such a plan should also take into consideration the (sectoral) functions given to the district within the broader framework of provincial development.

Conclusions

The regional problem in Indonesia is characterized by contrasts between densely populated 'Inner Indonesia' and thinly populated 'Outer Indonesia'. The great diversity in ethnic groups, cultures and religions, which are often regionally localized, are also thought to jeopardize the unity of the nation. Moreover, the 'New Order' that came to power in 1965 steadily legitimized tight control of the various socioeconomic groups. In short, there is not much space for participatory planning at present.

Therefore it is unsurprising that development planning is strongly centralized, hierarchical, sectoral and top down. Institutions that could theoretically counter centralism, such as the People's Representative Councils at the provincial and district levels and the Village Community Councils, have not played a crucial role in the decision-making process up to now. The BAPPEDAs (Regional Development Planning Boards) lack funds, skilled manpower and authority to achieve horizontal co-ordination of ministerial branches and local offices. They have no power to decide on the allocation of the regional Inpres funds.

On the one hand, the political reality may explain the centralized government structure and the limited devolution of decision-making powers to lower levels. Spatially and socially unbalanced development, as in the case of Sukabumi, however, seems to make a more prominent future role for the regions inevitable. From a development point of view, it is argued that decentralization of decision-making will increase the efficiency of development policy and will better suit the needs of the regions as well as their deprived groups. In this respect, several initiatives to strengthen the position of the BAPPEDAs were taken during the past decade, such as the Provincial Area Development Programme (PDP). Also, new planning procedures were introduced in 1981, offering potentially greater possibilities for bottom up planning.

However laudable the endeavour to decentralize development planning might be, it does not guarantee a more equal distribution of resources among social groups. The decisive case in point is to what extent regional authorities feel responsible for changing the position of poor groups in the regions. In this respect, it is both interesting and important to follow the efforts to decentralize development planning and to observe the actual operation of bottom up planning procedures in Indonesia.

With the Strategic Development Framework (SDF) approach, an example has been given of how, with limited resources, regional planning can enlarge the capacities of planning officers in the region and also include low-income groups in development planning. The SDF approach focuses on particular target groups, not only in terms of potentials but also deprivation. It stresses promising sectors but also backward areas. In this way it provides a tool for operationalizing the twin objectives of economic growth and redistribution of welfare. Moreover, it offers a framework for channelling the perceptions and aspirations of the identified target groups into the domain of public decision-making. It offers an alternative to development planners at lower levels where, largely as a result of the existing sociopolitical system, direct participation in planning within formal structures faces many constraints.

References

Booth, A. (1988) Central government funding of regional government development expenditures in Indonesia: past achievements and future prospects, *Prisma*, no. 45, pp. 7–12.

Geertz, C. (1963) *Agrarian Involution. The Processes of Ecological Change in Indonesia*, University of California Press, Berkeley, Calif.

Grijpstra, B. (1989) Western Java in development perspective (paper presented at the

International Workshop on Indonesian Studies, no. 4, Royal Institute of Linguistics and Anthropology, University of Leiden, 11–15 September).

Hardjono, J. (1983) Rural development in Indonesia: the 'top-down' approach, in D. Lea and D. Chaudhri (eds.) *Rural Development and the State*, Methuen, London, pp. 38–65.

Haskoning-Lidesco (1988a) *Bottom up Planning in Kabupaten and Kotamadya Sukabumi*, Directorate City and Regional Planning, Bandung.

Haskoning-Lidesco (1988b) *Medium Term Regional Development Plan 1989–1994. Development Region Sukabumi*, Directorate City and Regional Planning, Bandung.

Ichalsul Amal and Nasikun (1988) Decentralization and its prospects: lessons from PDP, *Prisma*, no. 45, pp. 23–32.

MacAndrews, C. (1986) The structure of government in Indonesia, in C. MacAndrews (ed.) *Central Government and Local Development in Indonesia*, Oxford University Press, Singapore, pp. 20–41.

MacAndrews, C., Fisher, H. and Sibero, A. (1982) Regional development in Indonesia: the evolution of a national policy, in C. MacAndrews and Chia Lin Sien (eds.) *Too Rapid Development. Perceptions and Perspectives from Southeast Asia*, Ohio University Press, Ohio, pp. 79–121.

MacDougall, J. A. (1982) Patterns of military control in the Indonesian higher central bureaucracy, *Indonesia*, no. 33, pp. 89–108.

Missen, G. J. (1972) *Viewpoint on Indonesia. A Geographical Study*, Nelson, Melbourne.

Otten, M. (1986) *Transmigrasi: Indonesian Resettlement Policy, 1965–1985. Myths and Realities*, International Work Group for Indigenous Affairs, Copenhagen.

Ravaillion, M. (1988) Inpres and inequality: a distributional perspective on the centre's regional disbursements, *Bulletin of Indonesian Economic Studies*, Vol. 24, no. 3, pp. 53–72.

Robison, R. (1978) Towards a class analysis of the Indonesian military bureaucratic state, *Indonesia*, no. 25, pp. 17–39.

Sanders, M. (1971) De regionale dimensie van de Indonesische ekonomische problematiek: een eerste oriëntatie, *KNAG-Geografisch Tijdschrift*, Vol. V, no. 2, pp. 98–107.

Schiller, J. (1988) Rural development strategy: learning from an Area Development Program, *Prisma*, no. 45, pp. 45–55.

Schulte Nordholt, N. G. (1987) *State-Citizen Relations in Suharto's Indonesia: Kawuli-Gusti*, Comparative Asian Studies Programme no. 16, Erasmus University, Rotterdam.

van den Ham, A. (1989a) An analysis of a Strategic Development Framework in Indonesia, *Regional Development Dialogue*, Vol. 10, no. 2, pp. 218–27.

van den Ham, A. (1989b) Regional development planning in West Java: the case of Sukabumi (paper presented at the International Workshop on Indonesian Studies no. 4, Royal Institute of Linguistics and Anthropology, University of Leiden, 11–15 September).

van den Ham, A. and Hariri Hady (1988) Planning and participation at lower levels in Indonesia, *Prisma*, no. 45, pp. 72–83.

Weinstock, J. A. (1989) Integrated rural development: the provincial area development program in West Java (paper presented at the International Workshop on Indonesian Studies no. 4, Royal Institute of Linguistics and Anthropology, University of Leiden, 11–15 September).

11

REGIONAL DEVELOPMENT PLANNING FOR RURAL DEVELOPMENT IN BOTSWANA

Ruud H. F. Jansen and Paul J. M. van Hoof

Introduction

For several decades, governments in developing countries have attempted to increase economic growth and diversification through direct and indirect intervention. The poor outcome of central planning and evolution of rural development policy have inspired various governments (e.g. Kenya, Indonesia and Botswana) to create new planning structures for rural development that allow for regional diversification and local authority in planning and decision-making.

Several authors have stressed the potential advantages of decentralized public administration and a regional development planning system for rural development (Rondinelli, 1981; Conyers, 1983; Smith, 1985; Sterkenburg, 1987; Conyers and Warren, 1988; Rondinelli, Cheema and Nellis, 1984; Rondinelli, McCullough and Johnson, 1989). A horizontal planning approach may facilitate the integration of sectoral planning. Decentralization may result in increased participation of the local population in planning and decision-making, thereby facilitating adjustment of policy measures and development activities relevant to local circumstances. This process would be enhanced by deployment of more detailed local knowledge by the planners working with the people in the respective areas. Regional decentralized development planning may therefore be more realistic, thus leading to improved rural development. Whether these potential advantages actually materialize depends on national and regional political and economic conditions, which determine the extent of the decentralization and the margins for government intervention (Slater, 1989).

Initially, rural development was equated with agricultural sector improvement.

During the 1970s the basic needs view was advocated and the World Bank (1975) wrote that

> rural development is a strategy designed to improve the economic and social life of the rural poor ... since rural development is intended to reduce poverty, it must be clearly designed to increase production and raise productivity ... the objectives of rural development therefore extend beyond any particular sector ... they encompass improved productivity, increased employment ... as well as minimum acceptable levels of food, shelter, education and health.

Integrated rural development approaches followed suit, stressing the need for multifaceted perspectives.

The role of land use planning in regional development has, for a long time, been rather obscure and was often deemed vague. Though it was acknowledged that land utilization was an important aspect within integrated rural development, natural resource studies and land use planning exercises were often either 'quick and dirty' or overdone. Only fairly recently has technical data-gathering been coupled with socioeconomic aspects of rural development, and land use planning and regional development planning have been reconciled and broadened to encompass common objectives for the improvement of the rural development process (Jansen, 1987, pp. 7-8; FAO, 1988).

Decentralized regional development planning for rural development in Botswana forms the subject of this chapter. More specifically, attention will be paid to the CFDA (Communal First Development Area) strategy – Botswana's version of integrated rural development – and improved land use planning (in the Southern District) as tools for appropriate rural and regional development. However, care is needed in drawing international comparisons as Botswana has certain characteristics that make development planning efforts difficult to imitate. Because Botswana is one of the few growth economies in Southern Africa, with a capital reserve in foreign exchange adequate to cover more than 3 years of imports, and a sustained annual GNP growth rate of approximately 12 per cent (World Bank, 1989), with only 1.5 million inhabitants and a long history of development planning and a relatively well-developed planning machinery, the content and reach of regional planning and rural development are substantial when compared to other developing countries.

By defining 'appropriate' within the limits indicated as potential advantages of decentralized public administration and regional development planning, this chapter aims to describe and analyse the CFDA strategy (through a comparison between North-East and Ngamiland Districts) and improved land use planning experiments in Southern District as tools for improved regional development planning and appropriate rural development. The framework for the two case studies is provided in the first two sections, which discuss decentralized development planning and trends in rural development respectively.

Major aspects of decentralized development planning

Nationwide development planning in Botswana is assigned to the Ministry of Finance and Development Planning (MFDP). This ministry is responsible for the

National Development Plans (NDPs), which outline national goals and priorities and cover a 5–6 year period. The Ministry of Local Government and Lands (MLGL) also plays a crucial role as the co-ordinating ministry for all local-government institutions and land use planning activities. The Rural Development Council (RDC), an interministerial committee, chaired by the Minister of Finance and Development Planning, is the highest government authority dealing with a wide range of rural development matters. MFDP houses the Rural Development Unit (RDU), which provides the secretariat for the RDC. It is in the RDU that the Communal Areas Co-ordinator, who has overall responsibility for the CFDA strategy, is based.

The ten districts in Botswana were created under British colonial rule in accordance with tribal boundaries. At independence, in 1966, the system of local government was partly changed. Today, the system of local government at district level comprises the Tribal Administration, District Council and Land Board. The Tribal Administration has as its main task the administration of justice under customary law. The District Councils, established only three months before independence, are elected bodies that have taken over many of the powers and the functions of the chief. Their statutory responsibilities include primary education, public health, social and community development, public water supplies and the construction and maintenance of most rural roads (Egner, 1987, p. 20). The Land Boards were set up in 1970 and have as their primary function the allocation of tribal land for various purposes, a prerogative previously held by the chiefs. The central government is represented by the District Commissioner, the District Officers and supporting staff; as mentioned before, all institutions come under MLGL's responsibility.

The District Development Committee (DDC) and the District Land Use Planning Unit (DLUPU) are important advisory committees. The DDC consists of central-government officers posted in the district and representatives of the District Council. Its main task is to co-ordinate the preparation of the District Development Plans (DDPs) and the implementation of development activities (Egner, 1978; Reilly, 1981). Annual plans are prepared with the aim of attaching budgetary allocations to projects agreed upon in the DDP. Through its key officials, the District Officer Development and the Council Planning Officer, the DDC is furthermore responsible for regular consultation with Village Development Committees and extension staff based in the rural areas so as to ensure the incorporation of grassroots ideas into the development process.

The Tribal Grazing Land Policy (TGLP), a far-reaching livestock development programme adopted in 1975, necessitated the establishment of a cadre of land use planners, the District Officers Lands (DOLs), who are nowadays assisted by the DLUPU. DLUPUs were intended to be more than land use advisory forums and were anticipated to become executive committees of planners functioning as a team in dealing with day-to-day development planning and implementation matters going beyond the strict realm of land utilization (Jansen, 1989).

The description above indicates that Botswana has paid ample attention to the decentralization of development planning and implementation in the post-independence period. It first created an institutional framework for economic planning, then delegated certain aspects of sectoral planning and implementation to the district level and subsequently added intersectoral co-ordination and elements of physical and land use planning to the responsibilities of district authorities (Sterkenburg, 1987, p. 166).

During the 1980s, however, evidence exists to suggest that a reversal of the government's decentralization policy took place. Increased control of local government budgets and abolition of the Local Government Tax in 1987 are examples of this development. It leads Gasper (1987, pp. 23–4) to the conclusion that

> the local government system is soundly designed, but was left short of resources. ...
> The implied administrative and centralist biases ... are obstacles to decentralisation.
> ... Together with any elite political opposition to devolution, they would help to explain
> why ... [in Egner's terms] ... administrative solutions to local authorities' problems
> have been preferred, rather than political solutions which give local authorities clearer
> authority.

In addition to the constraints already mentioned, the decentralized development-planning process has been hampered by discontinuity of staff at the district level. Lack of qualified staff and the lack of resources contribute to the fact that district planning has, in practice, not yet been fully integrated into the national planning system.

Symbolism and substance in rural development policy

Examining rural development policy in Botswana from the eve of independence in 1966 to date shows a growing realization of the needs and rights of the rural population on the part of the government. This has been most pronounced since the early 1970s. The emphasis has shifted progressively away from social service provision to a more integrated approach with particular attention to the creation of productive employment.

Though the objectives of rural development and the nature of strategies pursued have altered over the years, reflecting changing perspectives on the part of international donor organizations, two major aspects have played an important role in setting the direction of rural development, namely, the development of a growth economy based on mining activities and cattle industry and political acquiescence in the rural areas (Holm, 1982; Picard, 1987, pp. 231–32). An example of the realism of government policy was the priority given to attaining budgetary independence from Britain, which had paid most of the government budget during the 1960s. At that time, expenditure on construction of the capital of Gaborone and the development of prospective mining activities had been high. The coming on-stream of a copper-nickel mine in Selebi-Pikwe set the stage for changes in budgetary allocations and, ever since, revenues from mining (diamonds, copper, nickel, coal) have contributed substantially to the country's rapid economic growth.

The 1969 elections had shown that the ruling Botswana Democratic Party (BDP) did not have full control over the rural areas when Chief Bathoen II resigned from his chieftainship to join the Botswana National Front (BNF) opposition party, which then won three seats in southern constituencies. To pursue political acquiescence in the rural areas it became evident that the BDP would have to embark on an articulated rural development policy with visible and tangible results. A basic-needs-oriented programme, the Accelerated Rural Development Programme (ARDP), was initiated in November 1973, less than one year before the 1974 general election.

The ARDP was seen largely as a district-level activity, separated from central government operations (Picard, 1987, p. 238). It concentrated on providing social services and physical infrastructure to the rural population in the form of primary schools, health posts, road improvement and water reticulation. Judged on its own criteria the ARDP was a success in that many projects had been completed or were well under way by the election date. The ARDP increased capital investment in the rural areas more than fourfold during the 1973–6 period of its operation, spending more than 20 million Pula (Colclough and Fallon, 1983, p. 151). Roughly 50 per cent of this came from domestic development funds (Egner, 1978, p. 15) and benefited greatly from the earlier strengthening of district development staff and from strong political commitment (Colclough and McCarthy, 1980).

Holm (1982, p. 85) investigated the politics of rural development until the early 1980s, concluding 'that rural development is a relatively unimportant consideration in the voters' decision'. This particularly reflects the aftermath of ARDP or, rather, what had been held in store by the country's political leaders prior to the 1974 elections, as the major direction for rural development had been laid down in a government white paper in 1973. Entitled *National Policy for Rural Development*, this document clearly spelt out the future direction of rural development. It concentrated on a crucial land-reform measure, namely, the allocation of large tracts of tribal land to large cattle owners under exclusive tenure conditions (Government of Botswana, 1973, pp. 8–9). Incidentally, the policy also made mention of the fact that investing revenues in public services and infrastructure 'will not be enough' and that 'income earning opportunities must also be created'. Nevertheless, the emphasis on commercialization of the livestock industry pre-dominated and was claimed 'not necessarily [to] represent a threat to rural development' (*ibid.* p. 2). It was obvious that a major transformation of the rural areas would result from the pursuit of large scale commercialization of the livestock industry through the privatization of land. However, the voters were only made aware of this in July 1975, when the Tribal Grazing Land Policy (TGLP) was announced.

Though infrastructural developments would continue, the TGLP would have a large impact on rural development; in the words of Picard (1987, p. 243), 'If the ARDP represented symbolism and political quiescence, the TGLP represented substance and economic transformation'. At the heart of the policy was the allocation of huge commercial 'Texas style' ranches with the aim of alleviating grazing pressure in the overcrowded eastern grazing areas. The TGLP became a 'dismal failure' (Sterkenburg, 1987, p. 159). It did lead to the creation of several large ranches but it did not lead to better management, let alone curb overgrazing, especially in the communal areas. Instead of reducing inequalities in the rural areas, TGLP contributed to a further increase of them (Cliffe and Moorsom, 1979, pp. 46–9; Sandford, 1980; Bekure and Dyson-Hudson, 1982; Hinderink and Sterkenburg, 1987; Jansen 1987). The situation at the end of the 1970s was such that Holm (1982, p. 86) laments 'that Botswana's rural development programme receives minimal financing, shows relatively little concern with the mass of the population, is articulated by national rather than local officials, is heavily foreign funded and controlled, and yields few results which benefit the living conditions of the average voter'.

Within government circles, a realization came about some ten years ago that small, rural livestock owners and crop farmers were missing the boat completely. Consequently, efforts were also directed to subsidization programmes in the field of water development, herd management, ploughing and planting. The gradual change in rural development thinking from infrastructure and social service provision to productive employment creation is illustrated by the launching of a Financial Assistance Programme (FAP) for business promotion in the early 1980s that, amongst other things, has concentrated on various rural industries.

The evolution of the CFDA strategy in Botswana

So far, it has been shown that rural development in Botswana reflects a project and programme approach based on sectoral inputs in the provision of social services and the physical infrastructure. These programmes still exist today, and it seems justified to conclude that government budgets still favour urban areas, that rural development is still largely equated with services and infrastructure (Egner, 1987) and that sectoral inputs are being channelled into rural areas in a significantly top down manner. Notwithstanding this deduction, one government initiative that is trying to cut through these shortcomings has not yet been mentioned: the Communal First Development Area (CFDA) strategy. The CFDA idea is one of integrated rural development planning through participatory bottom up linkages between rural communities, district planners and central government. The development planning responsibility is further decentralized to the subdistrict level, whereas the decision-making responsibility remains at the district level. This strategy came about in 1980–81 as a strong reaction to the one-sided commercial focus of the TGLP to the neglect of the communal areas, and was expected to become the 'lead district level strategy through which sectoral rural development efforts are co-ordinated and focused to achieve maximum impact' (MFDP, 1985b, p. 6).

The CFDA policy involves the designation of a defined geographic area below the district level to receive priority in development, as it is considered impractical and impossible, within the limits of specialist support available from extension and district planning staff, to spread efforts over too wide an area. This special attention is limited to a specific period of time after which another area receives priority in development. Originally, a period of 3–5 years was deemed sufficient, but this proved to be too short in practice. In terms of population coverage, 5–20 per cent of a district's total population seems appropriate for CFDAs. In the CFDA, the development measures focus on increasing agricultural production and the expansion of non-farm employment. Increased participation will be achieved through a strengthening of local institutions and a more individual approach to the target group by planners and extension workers. The strategy includes as an essential element the preparation of a detailed land use and development plan based on the characteristics of the area, technical considerations of optimal resource use and needs expressed by the population. At the district level the CFDA working group, consisting of the senior district planners, is responsible for the CFDA planning and implementation. It is realized that this form of integrated rural development planning is a complex activity and can only be successful through co-operation between sectoral ministries at the district level. A CFDA co-ordinator will be (or has been) appointed to identify, implement and monitor ongoing projects (MFDP, 1982, 1985a, pp. 62, 86; Sterkenburg, 1987, p. 166).

Key elements of the CFDA strategy are:

1. *integration* of various programmes/policies of different sectoral ministries;
2. *decentralization* of implementation and management developments to the community level, and through this an increase in *participation*;
3. *adaptability* to differences in resources and conditions as they exist in these communities and their environments; and
4. *social justice* orientation by focusing on the poorer households.

The Rural Development Unit (RDU) of the Ministry of Finance and Development Planning is not very specific about the type of planning related to the CFDA strategy. It leaves room for the districts to adjust it to their specific needs. According to the RDU, CFDA planning should be a form of regional planning that has some elements of both a process approach and comprehensive resource-based planning. The emphasis on project identification and implementation seems obvious if we take into consideration the proposed scale of planning related to the CFDA strategy. Because the population per CFDA will be limited to approximately 20,000 people, the character of CFDA planning is closer to local planning than to intermediate planning.

Based on a socioeconomic profile of each CFDA, specific goals and objectives can be formulated that contribute to the main targets of agricultural-production increase and non-farm employment. These objectives in turn guide the selection of specific projects. The quality and quantity of information on the CFDA will improve over the life of the programme. It will be accumulated in a detailed land use and development plan that will guide further developments once attention has shifted to a second communal development area. Because the strategy is 'people oriented', in the sense that it is directed to the development of human potential, it is difficult to predict the effects and impacts of the projects. A good monitoring and evaluation system is needed to adjust objectives during and after the implementation and to guide the selection and planning of new projects (MFDP, 1982, pp. 9–41; Sterkenburg, this volume).

By choosing this planning approach, applicable to a small area and focused on specific objectives, the CFDA strategy can be more than just a concentration of efforts within district planning. If implemented as planned, it can be complementary to district planning, because

- it is people oriented, whereas district planning in Botswana is focused on social service and infrastructural planning;
- it is directed to project identification and implementation, whereas district planning is more directed to plan formulation;
- it creates room for a real dialogue between planning staff and the people in the CFDA, whereas participation in district planning hardly exceeds a pro-forma consultation exercise; and
- it is a true regional approach because it starts from an assessment of the development constraints and potentials of the area, which guide the selection of projects, whereas district planning has in practice thus far been a mere accumulation of sectoral plans.

CFDAs and rural development: a comparative analysis of the Ngamiland and North-East District CFDAs

How does the CFDA strategy operate in practice? Does it fulfil its function as a complementary planning exercise to district planning in Botswana? Though seemingly straightforward in principle, considerable variation can be seen in the implementation of this strategy across the ten districts. This is due to great differences in resource location and settlement patterns, to different attitudes of district planners towards this strategy and to the lack of guidelines and support from the central government (MFDP, 1985b).

Nine out of ten districts accepted the policy (South-East District considered itself too small) and selected a CFDA (Figure 11.1), but by 1988 only three districts had started a land use inventory and implemented special projects in their CFDA area. Ngamiland District is a good example of this group of districts. In the other districts, the implementation of the CFDA strategy proceeded slowly. No CFDA co-ordinators were assigned until recently, the planning efforts of the CFDA working group were not substantial and no special projects were identified and implemented. North-East District represents this latter group of districts.

CFDA Ngamiland District

This subsection is based on Sterkenburg (1987), pp. 168–79. Ngamiland District is situated in the extreme northwest of Botswana and is one of the biggest districts. It is characterized by a peripheral location, atypical natural resource conditions, a low population density and a highly vulnerable economy. Arable agriculture is limited to the fringes of the Okavango Delta, where most of the 75,000 inhabitants are concentrated. The remaining part of the district is most suitable for extensive ranching. Over 70 per cent of the population are engaged in the production of subsistence crops. Livestock rearing is more important than arable agriculture in terms of market value and income, but it benefits only a small proportion of the district's population. Other economic activities, such as fishing, hunting, crafts, commerce and services play only a marginal role in the district's economy. The dominant position of arable agriculture and livestock keeping under marginal

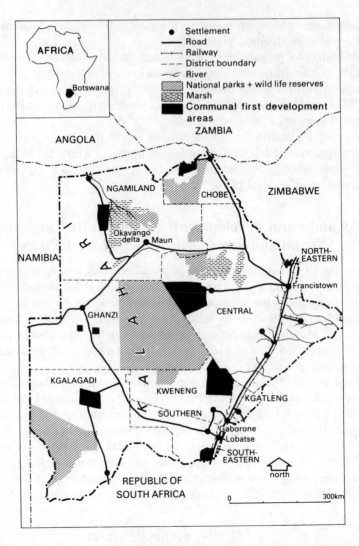

Figure 11.1 Botswana's Communal First Development Areas (*Source*: adapted from the Department of Survey and Lands: map of Botswana 1:1,500,000, 1982)

environmental conditions and at relatively low levels of technology led to a highly vulnerable economy. Prominent among Ngamiland's development problems is the district's isolated position, its low level of physical infrastructure and the related unfavourable marketing situation for livestock. In addition, it has rather vulnerable ecological conditions and limited natural resources of good quality. Increases in population and the commercialization of livestock led to a bigger herd, a varying degree of overgrazing and land use conflicts between agriculturalists and stock owners.

Ngamiland was the first district to select a CFDA, almost immediately after the proclamation of the strategy in 1981. The strategy fitted well into Ngamiland's ongoing development-planning activities. The district planning staff had already adopted land use planning as a major component of district development planning. This activity led to the demarcation of five production zones, one of which was selected as the CFDA. the selection of Nokaneng–Gomare–Etsha area was justified because of its high development potential, the existence of basic infrastructure and the representativeness for other communal areas in the district.

The planning exercise comprised several elements:

- The design of sectoral strategies by the CFDA working group, such as the expansion of Gomare as a district subcentre, the promotion of handicrafts and the improved marketing of agricultural produce.
- The execution of a socioeconomic base-line study and a farming-systems research project that resulted in gradual improvements to farming methods, an intensified extension service and, partly due to the recurrent drought, increasing attention to possibilities for small scale irrigation.
- The stimulation and improvement of the quality of non-agricultural activities, such as basketry, providing increased incomes for the CFDA population engaged in this activity.
- A land use exercise that resulted in an identification of the main land use problems and in area-specific proposals for improvement measures, specified by the view expressed by the communities involved.

Although the above-mentioned activities are extensive, especially if they are compared with those in other districts, the effects should not be overestimated. Successive years of drought during the 1980s have diminished the effects in terms of increased production, employment creation and income. Several factors contributed to the relative success of the CFDA planning activities in Ngamiland:

1. The proper selection of the CFDA, based on a division of the district into production zones and on available development potential in this area.
2. The availability of a land use and development plan for the CFDA before the identification and implementation of the projects began.
3. The early appointment of a CFDA co-ordinator in addition to high-quality district planning staff.
4. The presence of donor-financed and production oriented projects.
5. The establishment of a formalized management and reporting system supported by a continuity of staff and good working relations among them.
6. The designation of a subdistrict centre, combined with the construction of relevant infrastructure and the posting of staff, which shortened the distance between population and government.
7. The commitment of the District Council and politicians to the CFDA policy.

CFDA North-East District

North-East District is the second smallest district in Botswana, with around 45,000 inhabitants. It borders on Zimbabwe and is situated around Francistown, Botswana's second largest population centre, which is administratively separated from the district. The district comprises the area that Chief Lobengula granted to the Tati Company in perpetuity in 1888. The former residents of this area were allowed to stay, but were confined to a much smaller area known as the Tati African Reserve. These historical events still determine the present land use and development potential of the North-East District. Even today, more than 50 per cent of the land is privately owned freehold and is used for livestock keeping. The former reserve is heavily overcrowded, as 5,700 families live in an area with a carrying capacity for only 1,700 families with contemporary farming technique, and has a more scattered settlement pattern than found elsewhere in Botswana. Due to this overpopulation, agricultural productivity has stagnated, there is widespread outmigration of younger men to Francistown and other cities (78 per cent of the heads of household in the 15–44 age-group are female) and environmental problems of soil erosion and denudation are increasing. The cash economy depends heavily on remittances from migrant labourers and on government grants. Most people practise some kind of subsistence agriculture on small plots of 2–3 ha. Although the rainfall is good by Botswana standards (400–500 mm), soils tend to be overworked and have a low fertility, which results in low average yields, e.g. 200 kg/ha for sorghum. Cattle keeping is a privilege for only a few families, while most families have only some small stock. Due to the proximity of Francistown, which attracts all large- and medium-scale production activities and qualified labour, and to the scattered settlement pattern in the countryside, non-agricultural production is small scale, based on local needs and confined to only a few products such as dress-making, carpentry, bakery and brick moulding.

The selection of a CFDA in North-East District in 1982 resulted in a CFDA comprising two separate areas in the northeast and southeast part of the district in 1982. Because of the absence of a CFDA co-ordinator, confusion about the objectives and elements of the CFDA strategy, the lack of support from central government and absence of district planners, progress was slow until the end of 1984. At that time it became clear that there were too many conflicting development priorities in the district. There was competition from the development of Masunga as the new district headquarters and from the planning exercise concerning the four squatter areas in the district. Based on a recommendation of the CFDA working group, the District Council decided to redesignate the CFDA to the present Masunga area. In 1986, a local CFDA co-ordinator was appointed, who was joined by an expatriate in 1987. From that moment the CFDA co-ordinators have been trying to realize the non-specific targets as formulated by the CFDA working group:

1. To increase production in both agriculture and non-agricultural sectors, thereby diversifying the CFDA's economy.
2. To strengthen local institutions to enable them to play a leading role in local development.
3. To tackle some of the pressing resource management problems faced by the district, particularly those pertaining to land use planning, conservation and improved range management.

4. To improve social services and infrastructure.

Two kinds of activities are being executed to realize these targets. A detailed land use inventory is being compiled by a research team from the University of Utrecht in The Netherlands to assess and analyse the CFDA's present situation, its development constraints and its natural and human potential. Based on this information, a land use and development plan will be compiled and long term planning proposals will be designed and discussed with the CFDA communities and implemented after approval. While this inventory is ongoing, projects with a short term character that contribute to the realization of the CFDA objectives are being identified and implemented. They are based on existing local ideas and needs as expressed by the Village Development Committees (VDCs), which comprise chosen representatives of the village population and extension workers. Examples of such projects are the construction of several small- and medium-scale dams for the purpose of irrigation and watering cattle, several individual and group horticultural projects, the extension of the activities of a brigade (vocational training and production centre) and the execution of a rural industrial-promotion tour.

Though it is too early to evaluate these ongoing activities and to assess their effects on agricultural production increase and non-farm employment, the planning process to date can be analysed qualitatively by describing the factors and relations that influence the implementation of the CFDA strategy. Seen from the CFDA co-ordinators' point of view, the following parameters can be observed.

The human and natural resource position and development potential of the CFDA

More than in Ngamiland, the potential in North-East District CFDA is restricted. Increased agricultural production is most likely to be pursued through improvements to existing farming methods and through the introduction of small scale irrigation works for horticultural purposes. For these products, a small local market and possibly a market in Francistown are available. The extension of non-farming activities seems problematic. Due to strong competition from South African products on the already limited local market and the proximity of Francistown, where many people buy their daily needs, the local consumer market is limited. The market for home-industry products, such as sewing, knitting and brick moulding, is almost saturated. Development planners will have to take these limitations into consideration. Projects to encourage commercial activities will have to be small scale, labour intensive and initiated with financial support from the government if they are to be successful.

The interaction between the CFDA co-ordinator and the CFDA population

Participation in planning in Botswana is usually limited to a consultation tour for the preparation of the District Development Plan once in six years, or the Annual Plan. Instead of addressing the real needs and problems of the people, this consultation process usually results in a shopping list of wishes for more social services and infrastructure. Partly because of this kind of consultation, a 'dependency trap' has evolved in Botswana that led to the destruction of most 'self-help' activities. This creates an attitude that makes active involvement of the people in projects that will lead to sustainable development as proposed by the CFDA strategy increasingly difficult. However, this does not mean that the VDCs do not have any ideas on how employment can be encouraged. Interviews with VDCs showed that they do have constructive ideas, but do not know how to put these ideas into practice and are not aware of possible government support. Through intensive contact between CFDA co-ordinator and VDCs this situation is gradually improving.

The relationship between the CFDA co-ordinator and other government officials

In the framework for CFDA planning developed by the RDU no attention is given to the relation between the CFDA co-ordinator and extension workers. This omission is remarkable because there is a common interest or even a certain overlap in the work of the CFDA co-ordinator and that of Agricultural Demonstrator, Assistant Community Development Officer and Rural Industrial Officer. No guidelines for clear demarcation of their respective activities exist. This indistinctness is caused partly by the vague status of the CFDA co-ordinator and partly by the fact that the relevant ministries do not pay enough attention to the CFDA strategy as a guiding principle for planning and extension. The CFDA co-ordinators in North-East District try to minimize problems through extensive communication with the relevant extension workers.

A comparison of the CFDA activities in the two districts

One of the most important differences between Ngamiland and North-East District is the place of CFDA planning in district planning and the attitude of district planners towards the CFDA strategy. In North-East District, the CFDA strategy is not receiving the attention it should get from district planners. Because of this attitude, the CFDA working group, which should be the driving force behind CFDA developments, is merely a soundingboard for the CFDA co-ordinators. This lack of support could be alleviated if the CFDA co-ordinators had some authority. However, the bureaucratic position of CFDA co-ordinator is rather weak. As a local government officer working for the district council, which only provides services and infrastructure, he or she has only persuasive power over heads of departments and can be ignored by central government departments.

District planning in Botswana remains a blend of sectoral plans. In the North-East District Development Plan 3 (1986–9) the implications of CFDA planning on other sectoral plans are not covered at all. The top down and vertical planning in most ministries clearly conflicts with the more bottom up and horizontal approach in CFDA planning. In the absence of an Annual Plan in North-East District and a clear monitoring system, the task of the CFDA co-ordinator becomes even more problematic because he or she has no overview over the planned activities of the sectoral ministries in the area. More than once, projects have been implemented without the officer's knowledge.

The description of CFDA planning in Ngamiland and North-East District shows that a combination of:

- a good selection procedure for the CFDA, based on a land use evaluation of the district and a ready available land use and development plan for the CFDA at its start, together with
- committed district planning staff, who can attract external funds and support, and
- a co-operative and guiding central government,

seems to be essential for successful implementation of the CFDA strategy.

Reduction of the problems concerning the implementation of the CFDA strategy must start at central-governmental level. Several major and minor adjustments are possible in order to improve the contemporary situation, such as

- improving the status of the CFDA co-ordinator, by placing him or her under the District Administration or by giving him or her a small development budget of his or her own;
- clarifying the tasks of district planners and extension workers with respect to the CFDA strategy;
- upgrading this horizontal and 'grassroots' planning approach in relation to sectoral approaches; and
- concentrating CFDA efforts on the creation of productive activities would be helpful in reducing the confusion concerning the CFDA strategy at district level.

Towards improved land use planning for rural development: a case study of Southern District

Land use planning in Botswana started with the launching of the Tribal Grazing Land Policy (TGLP), which necessitated the zoning of districts into land utilization areas based on principal existing uses and desirable future developments. Subsequently, district-level land use planning has taken on board several other tasks requiring the input of a geographer/cartographer, such as village planning and agricultural planning. These duties were later to be shared with the District Physical Planner(s) and the Land Use Officer(s). During the last decade, land use planning has evolved from a pure and straightforward technical land use zoning exercise to a more comprehensive contribution to rural (economic) development planning. The manner in which this occurred will be illustrated with reference to the Southern District.

However, in order to understand the land use planning procedures fully, it is necessary to outline the various kinds of land tenure existing in the country. There are three legal categories of land: state land, tribal land and freehold land.

State land, defined by the post-independence State Land Act, comprises approximately one fifth of the total area of the country, including, for example, the national parks, game reserves and most of the township areas. Applications for land rights are handled by the Department of Surveys and Lands (DSL) and title deeds are registered by the Attorney General's Chambers. Land use planning for state-land areas in the districts is one of the responsibilities of the district administration, in particular the DOL, especially regarding technical investigations for the utilization of the state land areas. The MLGL and DSL, however, retain overall responsibility for decision-making, implementation and land administration in state land areas.

The largest proportion of land is tribal land, which comprises roughly 75 per cent of the country's total area. Land tenure in tribal areas has its origin in customary law and is regulated by the Tribal Land Act. A customary grant of land is made by the Land Board to members of the tribe for residential, grazing and ploughing purposes only. The use of residential and ploughing land is exclusive whereas grazing land is used communally. In addition to allocations under customary law, the Land Board has the right to allocate land under common-law tenure to any person for whatever purpose by way of short or long term leases or even ownership with the written consent of the MLGL. Land use planning in tribal areas falls within the portfolio of the Land Boards through the acceptance and implementation of land use plans prepared by the District Land Use Planning Units (DLUPUs). As far as political aspects are concerned, Land Boards are obliged to consult District Councils for consent. This is also a moral duty as part of the consensus approach to the approval of local authority plans. However, ultimate responsibility to the MLGL remains with the Land Board as an independent local government authority.

Freehold land is the smallest category of land in the country: only 6 per cent of the total area. It is mainly found in the farm blocks around Francistown and to a lesser extent around Gaborone. The origin of freehold land is precolonial, when chiefs were still entitled to sell land to private companies. The tenure system is one of absolute ownership of land and was created when the state permanently transferred its title to a private individual, natural or juridical. The title does not lapse with time and is inheritable, registrable, transferable and mortgageable. Land use planning occurs mainly in the urban freehold areas (e.g. in part of the Gaborone Planning Area); in rural areas, land-utilization monitoring and land use legislation form the major areas of attention.

Land use planning can be defined as a means of helping decision-makers decide how to use the land by evaluating land and alternative patterns of land use systematically, choosing the use that meets the specified goals and then drawing up policies and programmes for the use of the land (Dent, 1988). In Botswana, land use zoning plans have been prepared as an integral part of the TGLP land use planning exercise. However, zoning is only part of the overall exercise. Its main purpose is to control land use; the zones should be designed to regulate potentially conflicting land uses and to prohibit undesirable development (Jansen, 1987, p. 35). If seen as a planning tool, land use planning exercises should therefore not end at the stage of indicating the best land uses but should include all projects, programmes and policies to be implemented by those involved in using the land so as to achieve the desired land utilization.

Following the approval of the 1986 Land Use (Zoning) Plan in Southern District (Figure 11.2), the district planners embarked on the conduct of a district-wide

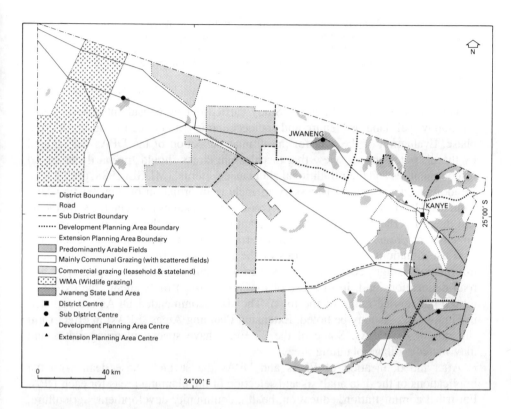

Figure 11.2 Southern District land use situation and proposed planning areas
(*Source:* adapted from MLGL/Southern District, 1988)

survey with the aim of selecting Development Planning Areas (DPAs). The DPAs
would encourage more accurate plan formulation and facilitate plan implementation
because

1. closer planning attention to grassroots problems occurring across groups of
   villages/communities would become more feasible; and
2. manageable areas for better execution of government policies and programmes
   would be created as well as a better structure for extension services and
   supervision of field projects (Jansen, 1987, p. 16; MLGL, 1988, pp. 1–2).

The initiative can be closely linked to experiences with comprehensive planning
to that date and the implementation of the CFDA strategy. The latter, as has been
mentioned, was most successful in Ngamiland District, where the area selection
had been based on 'production zones' identified within the district. These production
zones closely resemble the DPAs anticipated in the Southern District Planning
Study and the fundamental idea is one of co-ordination and integration of sectoral
rural development programmes in delimited geographic areas (MLGL, 1988, pp.
1–2). Ideally the CFDA should therefore coincide with one of the selected DPAs.

The deviation from clear-cut land use planning exercises may be highlighted
by the explanation that

the concept of planning areas should not be confused with land use zones ... the idea of [the selection of] planning areas is essentially to disaggregate the district into parts which have a common development strategy or to which such a strategy could be applied ... it is to some extent unavoidable that a planning area will consist of more than one land use zone.

(Jansen, 1987, pp. 16–17)

The Southern District Planning Study carried out by consultants concentrated on biophysical, land use/tenure and socioeconomic data collection during its first phase. Evaluation of the data for the ultimate selection of the DPAs was guided by the district planners' view that the planning areas should 'primarily be based on communities and their traditional allegiance patterns' (MLGL, 1988, pp. 11–13). In addition, it was found necessary to include information on the existing land tenure and land use situation, partly because of the linkages between the settlements and their hinterlands and no less significantly because of the existence of specific government programmes and policies pertaining to the different land tenure and land use situations. The boundaries of the DPAs were fine tuned by using such information as agro-ecological zones, census-enumeration areas, topography and road delineations and, in the case of the mining town of Jwaneng, the extent of the 20-km service range of the town. The six recommended DPAs are depicted in Figure 11.2. As may be noted, Extension Planning Areas (EPAs) in effect form subunits of the DPAs. Some of the 22 EPAs have special planning status and may include village planning areas.

After the delineation of EPAs and DPAs the second phase dealt with the implications of the data analysis and selection of the planning areas for each DPA. For tribal administration, education, health, community development, agriculture, Land Board and Council (constituencies), the impacts were listed and measures proposed to align planning boundaries with existing ones. Improvements of services were indicated and recommended based on

1. an analysis of existing situations;
2. a comparison of resource allocation between planning areas; and
3. the application of national minimum service and basic needs guidelines, so as to standardize facilities in each DPA.

Although it is still too early to evaluate the effects of rural development as now being pursued in Southern District, it seems fair to conclude that the approach is potentially successful. Compared with other districts, the Southern District methodology is certainly more comprehensive, integrative and following a clear direction. It should, however, not be misunderstood as being the end of development planning. Rather the DPAs should promote better plan and programme formulation and execution of existing government schemes. There is no longer any justification for the continued existence of the erstwhile bias towards *ad hoc* planning. The new planning approach shows how principles of land use planning combined with ideas on 'planning for and with the people' can lead to an integrated planning framework.

Conclusion

This chapter has described and evaluated, in a qualitative manner, two 'tools' for regional development planning in Botswana. Both of them seem to be appropriate to further improvement of regional development planning and to the improvement of the district development planning system in such a way that the theoretical advantages of decentralized planning, i.e. improved sectoral integration, participation and plan formulation, can be better achieved.

The CFDA strategy contains several elements that make it complementary to district development planning. This feature is essential to further improvement of regional planning for rural development in Botswana, the content of which is becoming increasingly complex. These elements are its process approach combined with an intensive consultation process for short term project selection and implementation, thus giving planning a more grassroots character. The horizontal planning approach in which land use planning is integrated will guide long term project selection and implementation.

The Southern District improved land use planning experiment also contains elements for better regional planning. Its land use planning methodology is more comprehensive, integrative and directed compared to previous *ad hoc* land use planning exercises and those still prevalent in other districts. As such, it provides a framework for improved district development planning and consequently contributes to the achievement of the potential advantages of decentralized development planning. A division of the district into District Planning Areas (DPAs) can promote better plan and programme formulation if coupled to sectoral integration. The execution of existing government schemes can be facilitated through the alignment of the various extension area boundaries to coincide with the DPA boundaries in the district. In addition, improved selection of CFDAs and subsequent development areas can be pursued.

Nevertheless, implementation of the total programme to date has been constrained by a lack of commitment by politicians and bureaucrats at the national level. However, the example of Ngamiland District indicates that the CFDA strategy can be implemented successfully through proper land use and plans for rural development, supported by dedicated district officials and politicians. The potentially advantageous Southern District land use planning exercise operates within the same national and district planning environment. Therefore, a similar reservation about its successful implementation seems to be justified.

The present and future extent of the government's commitment to rural development will be the primary determinant of whether these two appropriate, and mutually supportive, tools for more effective regional development planning will contribute constructively to improving the situation of the rural poor.

Botswana's continued economic buoyancy, which is well-nigh unique in Africa if not in the Third World, means that financial stringency, the dictates of IMF structural adjustment policies and related pressures from foreign donors, as experienced in many other developing countries, do not threaten regional development policy. Far more than elsewhere, the prospects for territorial planning depend on domestic factors, and the Botswana experience ought therefore to be of continuing international interest.

References

Bekure, S. and Dyson-Hudson, N. (1982) *The Operation and Viability of the Botswana Second Livestock Development Project (1497–BT): Selected Issues*, Ministry of Agriculture, Gaborone.

Cliffe, L. and Moorsom, R. (1979) Rural class formation and ecological collapse in Botswana, *Review of African Political Economy*, no. 15/16, pp. 35–52.

Colclough, C. and Fallon, P. (1983) Rural poverty in Botswana: dimensions, causes and constraints, in Dharan Ghai and Samir Radwan (eds.) *Agrarian Policies and Rural Poverty in Africa*, ILO, Geneva, pp. 129–53.

Colclough, C. and McCarthy, S. (1980) *The Political Economy of Botswana: A Study of Growth and Distribution*, Oxford University Press.

Conyers, D. (1983) Decentralization: the latest fashion in development administration?, *Public Administration and Development*, Vol. 3, pp. 97–107.

Conyers, D. and Warren, D. M. (1988) The role of integrated rural development projects in developing local institutional capacity, *Manchester Papers on Development*, Vol. 9, pp. 221–6.

Dent, D. (1988) *Guidelines for Land Use Planning* (fifth draft), FAO, Rome.

Egner, B. (1978) *District Development in Botswana* (report to SIDA), Gaborone.

Egner, B. (1987) *The District Councils and Decentralisation 1978–1986* (report to SIDA), Gaborone.

FAO (1988) *Land Evaluation and Farming Systems Analysis for Land Use Planning* (first draft) (mimeo).

Gasper, D. (1987) *Development Planning and Decentralisation in Botswana* (second draft paper) (mimeo) Institute of Social Studies, The Hague.

Government of Botswana (1973) *National Policy for Rural Development* (government paper no. 2), Gaborone.

Hinderink, J. and Sterkenburg, J. J. (1987) *Agricultural Commercialisation and Government Policy in Africa*, Routledge & Kegan Paul, London.

Holm, J. (1982) Liberal democracy and rural development in Botswana, *African Studies Review*, Vol. 25, no. 1, pp. 83–102.

Hoof, van P. J. M. and Jansen, R. (1989) North-East District CFDA: Demography, farming systems and employment inventories for land use planning (technical report Vol. 3), University of Utrecht.

Jansen, R. (1987) *Towards Integrated Rural Development Planning in Southern District, Kanye*, Government of Botswana, Kanye.

Jansen, R. (1989) *Environmental Planning in Botswana: Principles, Practices and Proposals* (report to Swedeplan), Gaborone/Utrecht.

Ministry of Finance and Development Planning (MFDP) (1982) *Communal First Development Areas as a Strategy for Rural Development*, Government Printer, Gaborone.

Ministry of Finance and Development Planning (MFDP) (1985a) *National Development Plan*, Government Printer, Gaborone.

Ministry of Finance and Development Planning (MFDP) (1985b) *Second Progress Report and Recommendations CFDA Strategy*, Government Printer, Gaborone.

Ministry of Local Government and Lands (MLGL)/Southern District (1988) *Southern District Planning Study* (final report), Environmental Consultants, Gaborone.

Picard, L. A. (1987) *The Politics of Development in Botswana. A Model for Success?*, Lynn Rienner, London.

Reilly, W. (1981) District development planning in Botswana, *Manchester Papers on Development*, no. 3, University of Manchester.

Reilly, W. (1983) Decentralisation in Botswana – myth or reality?, in P. Mahwood (ed.) *Local Government in the Third World, the Experience of Tropical Africa*, Wiley,

Chichester, pp. 28–69.

Rondinelli, D. A. (1981) Government decentralization in comparative perspective: theory and practice in developing countries, *International Review of Administrative Sciences*, Vol. 47, pp. 133–45.

Rondinelli, D. A., Cheema, G. S. and Nellis, J. R. (1984) Decentralization in developing countries, a review of recent experience, *World Bank Staff Working Papers*, no. 581, Washington, DC.

Rondinelli, D. A., McCullough, J. S. and Johnson, R. W. (1989) Analysing decentralization policies in developing countries: a political-economy framework, *Development and Change*, Vol. 20, no. 1, pp. 57–87.

Sandford, S. (1980) Keeping an eye on TGLP, *NIR Working Paper 31*, Gaborone.

Slater, D. (1989) Territorial power and the peripheral state: the issue of decentralization, *Development and Change*, Vol. 20, no. 4, pp. 501–31.

Smith, B. C. (1985) *Decentralization: The Territorial Dimension of the State*, Allen & Unwin, London.

Sterkenburg, J. J. (1987) *Rural Development and Rural Development Policies: Cases from Africa and Asia*, Elinkwijk, Utrecht.

World Bank (1975) *Rural Development Sector Policy Paper*, Washington, DC.

World Bank (1989) *World Development Report*, Oxford University Press.

12

STRUCTURAL ADJUSTMENT, DECENTRALIZATION AND EDUCATIONAL PLANNING IN GHANA

Bill Gould

Structural adjustment in Ghana

The almost universally dismal economic performance of Sub-Saharan Africa (SSA) as a whole and most individual countries of it in the last 20 years has been a matter of profound disappointment and continuing concern for all students of African development. Among them have been many geographers for whom the development crisis has inevitably brought into their work more explicit concern for macro-economic policies and processes than had hitherto been the case, for it has been to macro-economic factors associated with what are generally known as structural adjustment policies, and associated particularly with the World Bank and the IMF that the focus of so much attention in development studies has been and is being directed. Structural adjustment is seen by these powerful agencies and others, such as the European Community (Stevens and Killick, 1989) and by many governments of Third World countries, with varying levels of commitment and enthusiasm, as a path, if not *the* path, out of the apparent development impasse.

The range of fiscal, monetary and employment measures associated with structural adjustment in Africa were first elaborated in the World Bank's *Accelerated Development in Sub-Saharan Africa: An Agenda for Action* (the Berg Report, 1981) and confirmed and extended in a stream of subsequent reports (e.g. World Bank, 1984b, 1989). Stabilization of the currency, liberalization of trade and rolling back the involvement of the state in social as well as economic activities, it is argued, are necessary for long term economic recovery. These are to be achieved by a range of macro-economic measures associated with the 'New Right', including vast reductions in government expenditures, switching production to export-oriented activities in agriculture and in manufacturing and removal of controls on foreign trade and domestic prices. It has proved rather less difficult to reduce government expenditures, both recurrent and more particularly capital expenditures, than it has been to stimulate production and trade. The level of demand in African economies has been sharply reduced but the supply side has not been sufficiently stimulated, so that the net effect has been severe deflation of the economy. The theory and practice of structural adjustment policies have been subject to considerable critique both from academic sources (e.g. Allison and Green, 1983) and in governmental and UN documents (e.g. UN Economic Commission for Africa's *African Alternative Framework*, 1989). These argue, in particular, that the policies invoked to achieve the objectives of structural adjustment create such severe hardship in the short term on the incomes and quality of life of the populations affected, that the cure may be more harmful than the illness. Individual countries have experienced rather different effects of structural adjustment programmes and, despite political reluctance, most have found it preferable to be involved in them in order to receive loans and other economic support from the major multilateral agencies (Colclough and Green, 1988).

Ghana is probably the most widely discussed African country affected by structural adjustment programmes. A massive injection of external resources has been required to support the programmes, amounting to over US $1,000 million over the period 1983–8 (Green, 1988, p. 12; World Bank, 1984a). In the 1950s and leading up to independence in 1957 Ghana was the richest country in SSA (excluding South Africa and Namibia), but since the mid-1960s economic performance has fallen steadily to the extent that GNP per capita in the late 1980s was in real terms much the same as in 1950. Real per capita GDP, indexed at 100 in 1950, rose to a peak of 194 in 1962, but had fallen to a low of 86 for 1983, recovering since then. Per capita income declined by over 2 per cent per annum during the 1970s and 1980s. Recurring political problems, internal economic mismanagement and declining external conditions (particularly in the price of cocoa) all contributed to massive economic collapse. The government of Ghana itself admits that in 1981 'Ghana was illiquid, virtually bankrupt' (Ghana Government, 1987, p. 2). The country has been dependent upon structural adjustment loans and conditions surrounding them since 1983, to the extent that Ghana is frequently seen as a test case for World Bank free-market policies, and with some signs of success. Government budget deficits have been converted into surpluses, mainly by reducing the civil service payroll. Inflation levels have fallen, and the currency is now freely convertible, albeit at a falling rate. The government has put 32 state-controlled companies up for sale, and per capita food production has recovered from its 1983 low to levels of the 1970s (Moseley and Smith, 1989; World Bank, 1989).

Even stern critics of the general rationale for the World Bank's approach, such as Green (1988), attribute some success to the policies in the macro-economic sphere, but wonder whether it is possible to have 'adjustment with a human face', in the telling phrase coined by Green and others in a study for UNICEF (Cornia, Jolly and Stewart, 1987), to integrate policies for social improvement into the macro-economic matrix. Specifically in Ghana, as part of the strategy of the

Provisional National Defence Council (PNDC), in power since a *coup d'état* in 1981 under Flight Lieutenant Jerry Rawlings, there has been an Economic Recovery Programme (ERP), as the package of structural adjustment measures is called by the Ghanaian government. It seeks to effect macro-economic policies to enable internal economic change, to generate public and private revenues that will permit improvements in education, health and other social sectors, benefits that can be enjoyed by the population as a whole and not only by the direct beneficiaries of improvements in economic production and higher incomes.

This chapter considers decentralization in the Ghanaian context of macro-economic change – one of the several policies implemented as part of the ERP to spread its benefits in space, outwards to rural areas and downwards from the capital to small rural communities. In particular it will consider the role of the education sector in the decentralization programme, and the ways in which power relationships and technical responsibilities within education are being altered through it. In the following section the spatial structures of Ghana, both at the rural–urban and regional scales, are explored to set a context for analysis in separate sections of the general justification for decentralization and for consideration of the specific circumstances of education provision and changes that will be necessary if the decentralization policies are to be fully and successfully implemented.

Rural–urban and regional imbalances

One of the major changes in the geography of Africa in the 1980s has been what Jamal and Weeks (1988) have called 'the vanishing rural–urban gap'. Their ILO study of household incomes in a range of countries identified a near-universal trend of a narrowing gap, more as a result of falling urban incomes than as a result of rising rural incomes, and in part due to the effects of structural adjustment policies. These have sought, on the one hand, to raise producer prices to farmers and, in Ghana, this has primarily meant cocoa prices. Though much less important than in the 1950s and 1960s, cocoa still earns more than 50 per cent of Ghana's export revenues, and the prices paid to farmers by the Cocoa Marketing Board in Ghana have risen substantially since 1984. However, at the same time, policies have sought to hold down urban wages, especially in the large public sector. In Ghana there have been many redundancies and higher levels of unemployment as a result of reduced bureaucracies in line with the government's commitment to prune the public payroll by 5 per cent per annum. Gross rural–urban income differentials have narrowed, yet urban migration has not lessened. This is in direct contradiction of neoclassical explanations of migration that see the rate of rural–urban migration as a function of urban wages and employment, of demand for urban workers rather than the supply of potential migrants (e.g. Todaro, 1976). Jamal and Weeks (1988) explore (what most geographers have regularly considered in migration studies) the way that the functional and intra-household complementarity between rural and urban areas and associated flows of income and other goods sustain, indeed may intensify, the migration flows in time of economic crisis (Gould, 1988b; Potter and Unwin, 1989). Jamal and Weeks conclude that

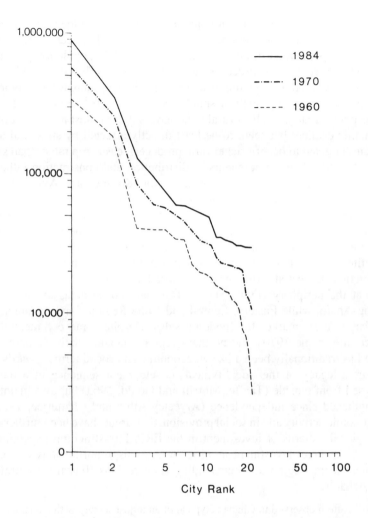

Figure 12.1 Ghana: city size distribution, 1960, 1970, 1984

the social distribution of income is generally moving adversely in Africa, a conclusion that is consistent with, indeed expected by, studies of the effects of structural adjustment in action. In rural and in urban areas, rich and poor households and individuals are increasingly differentiated.

Though Ghana was not explicitly studied in the ILO review, there is strong evidence to suggest that Ghana's experience confirms the general findings. Urban incomes have certainly fallen in real terms. Civil servants' wages are now very low and generally insufficient to maintain a family on the formal salary alone (Ewusi, 1987). Furthermore, many former civil servants are no longer employed in public service but many are still in urban areas seeking an income in the informal sector (Jonah, 1987). Producer prices, notably for cocoa, have risen, though they still lag behind real prices obtained in the boom cocoa years of the

1950s and early 1960s. The rural–urban income gap in Ghana has narrowed, yet migration to towns continues at a rapid pace. In 1960 the census recorded 26 per cent of the population in urban centres of over 2,000 population; that proportion was 35 per cent as recorded by the 1984 census. The 1960, 1970 and 1984 censuses recorded a very similar pattern of primacy, though with some evidence of rather more rapid growth of smaller centres by 1984 (Figure 12.1). Overall it is the poor, urban as well as rural, who have suffered most as a result of previous economic decline, but seem to be least directly affected by structural adjustment policies targeted to benefit richer rural producers, cash crop rather than subsistence farmers, and urban entrepreneurs in distribution and construction rather than the mass of poor migrants in the informal sector (Kasanga and Avis, 1988).

The effects of economic decline and macro-economic policy responses to it at a regional scale have been much less studied. In general, policies have favoured commercial production in the rural sector so that it is the most commercialized agricultural regions, cash-crop producing areas, that are more likely to have benefited rather than relatively backward regions dominated by subsistence production. In spatial terms these are more likely to be in or near the core rather than at the periphery (Figure 12.2). The main cocoa areas are in Ashanti and Brong-Ahafo, while Eastern, Central and Volta Regions have favoured access to the large Accra market for foodcrop sales. Despite some commercialization of agriculture in the 1970s, the northern regions remain relatively poor. In Ghana there has traditionally been a bias in economic and social terms towards the south, in part a legacy of the 'ideal-typical' development sequence of a development imposed from outside (Taaffe, Morrill and Gould, 1963) but also maintained and strengthened since independence (Aryeetey-Attoh and Chatterjee, 1988). Biases in economic activity and in social provision, they argue, have been further reinforced by regional patterns of investment in the ERP. Infrastructural expenditure, much larger than direct investment in agriculture, is disproportionately concentrated in Greater Accra Region and surrounding rural regions (Eastern, Central, Western and Ashanti):

> This pattern observed in Ghana is typical of an unjust society with a distinction between a socially deprived periphery (Northern and Upper Regions), the intermediate regions (Ashanti, Brong-Ahafo, Western and Volta Regions) and the affluent Greater Accra Capital district. The social injustice is a consequence of investment policies – industrial, infrastructural and social – and of high rewards for those individual capitalists or entrepreneurs, managers or professionals able to compete in a mixed economy largely controlled by the minority or elite in Accra
>
> (*Ibid.* p. 36)

In regional terms, then, the established patterns of economic activity and social provision seem to reinforce the current biases in the political economy of the country that basic needs strategies or even structural adjustment with a human face might seek to address.

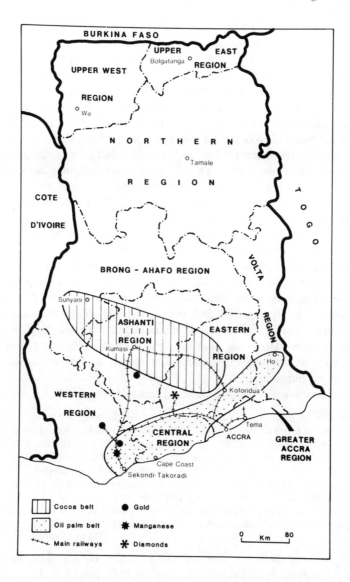

Figure 12.2 Ghana: administrative regions and major economic activities

Decentralization and the Economic Recovery Programme

Ghana's Economic Recovery Programme has proceeded in two distinct phases: ERP I, 1983–6, had economic stabilization as its first aim, halting the economic decline of the previous years; and ERP II, 1987–9, sought to embark on a programme of growth and development, with special emphasis on the social services. This involved change and rehabilitation in education, health, housing, water supply, culture and other sectors, as well as parallel changes in economic sectors, gathered together in 1988 for the implementation of a co-ordinated programme, PAMSCAD (Program of Action to Mitigate the Social Costs of Adjustment) (Ghana Government, 1988, p. 15). PAMSCAD comprises 23 'socially conscious' projects over a 2–3-year period, financed through the UN Development Programme (UNDP) and several bilateral agencies to the value of US $100 million. The largest component (over 40 per cent of the budget) is for employment generation, and about 20 per cent is allocated to 'redeployment', promoting the economic adjustment of retrenched civil servants. Basic needs and education provision comprise most of the remaining projects (*West Africa*, 11 January 1988). More generally, resources in the ERP are allocated to infrastructure and social sectors, the role of the state in directly productive activities being weakened in preference to encouragement of the private sector.

The allocation to the social sectors within the ERP reflects the broader objectives of the structural adjustment programme of weakening the role of the bureaucratic state. Part of the ideology of structural adjustment is that the state 'distorts' the natural workings of the market and thereby creates inefficiencies that can only be reduced by rolling back its economic power. So too, it is argued, in the social sphere: the state is less able to provide an adequate and appropriate range and quality of services than are those who consume the services. Social provision, traditionally centrally planned and supply led, should be demand led and the responsibility of local communities, organized round their own felt needs and priorities. A decentralized social structure is therefore consistent with the general range of structural adjustment measures, and further undermines the strength of the centralized bureaucracy.

Decentralization was given a political rationale in the ERP:

> Sustained work at both economic and political levels will be required since success in establishing new democratic structures will ultimately be founded on our overall capacity in the coming years to satisfy our basic needs of food, shelter, clothing, etc. An important element in the Government's planned and co-ordinated programme of economic and social development is the decentralization programme. It is not merely a 'political' programme to ensure grass roots participation in decision making. Rather it is an expression of a fundamental principle that it is those who participate in the productivity of society who can also lay claim to participation in the political institutions through which society determines its course and takes decisions for its well being.
>
> (Ghana Government, 1987, p. 21)

Accordingly, new districts were created in 1988. By the time of the District Assembly Elections (beginning in December 1988 and extending into February 1989), 110 districts had been created (Table 12.1). At the time of writing (November 1989), no map is yet publicly available to delineate these districts, and there is continuing dispute in many areas over precise delineation of district boundaries. Each district serves just over 100,000 people on average and, with the obvious exception of Accra, the regional range is fairly narrow (87,000 per district to 140,000 per district). As would be expected, there is a fairly close correlation with population density: the districts of low-density regions (Northern, Upper West) have relatively small populations; those of high-density rural regions (Ashanti, Eastern) have large populations, with Central standing out as rather overprovided with respect to its high population density. The special case of

Table 12.1   Districts and population by region

Region	Population (1987 est.)	Population density (per km^2)	Districts (Dec. 1988)	Population/ district	Pupil: teacher ratio (1987–8)
Ashanti	2,284,710	86	18	126,928	25:1
Eastern	1,815,359	87	15	121,024	24:1
Greater Accra	1,618,448	441	5	323,690	31:1
Volta	1,293,410	59	11	117,583	19:1
Brong-Ahafo	1,347,358	31	12	112,280	23:1
Northern	1,305,548	17	15	87,037	18:1
Western	1,282,620	48	11	116,602	24:1
Central	1,220,580	116	12	101,715	27:1
Upper East	842,942	87	6	140,490	20:1
Upper West	476,094	24	5	95,219	28:1
Total	13,487,069	52	110	122,610	23:1

(*Source:* Various tables in Pandit and Asiamah, 1988.)

Greater Accra, with a population per district nearly three times the national mean, is not unexpected.

District Assemblies are the highest political authority in each district. They comprise two-thirds elected members, one-third nominated members and the District Secretary. They play a central role in managing services and prioritizing resource allocations:

a) improving the access of the people to productive and social capital in the district;
b) increasing the availability of such capital in the district;
c) taking appropriate measures and embarking on projects and programmes that would increase the yield of physical and human capital; and
d) encouraging and/or undertaking such activities as will eliminate bottlenecks to maximum production and, specifically, seeking better balance and stronger linkages between the various economic sectors in the district.

(Ghana Government, 1987, p. 22)

Districts have, under PNDC laws, power to raise revenue and to make loans to 'break any stranglehold on production that may be posed by the system of land tenure with minimal social disruption' (Ghana Government, 1987, p. 23). In the social sector they have responsibilities for basic education (primary and secondary schools) and primary health care. They are also responsible for water supply and public health, for soil and water conservation and for the maintenance of all public roads other than trunk roads. Each assembly must have four sector committees: economic development, justice and security, technical infrastructure and social services. They may raise revenues locally by taxing marketed foodstuffs originating in the district, and so are likely to designate or build formal markets. It is likely that the districts will be subject to what is known in Ghana as composite budgeting, i.e. that each district will be given a global sum to be allocated within certain limits, to the various sectors over which it exercises responsibility. This will replace the strictly sectoral budgeting that has hitherto been the practice for regional administrations.

The political objective of establishing a decentralized participatory democracy is not new, but its operationalization has been given urgency by the objectives of the ERP and support from funds made available through ERP. In particular, the programme of UNDP, 1987–91, has explicitly sought, as one of its three major strategies, to promote grass-roots participatory rural development (UNDP, 1987). UNDP is essentially a co-ordinating institution, acting as an implementational umbrella organization for the multisectoral activities of the many UN agencies. As such it ensures that UN funds are channelled consistently and are compatible with the objectives of government. In Ghana it has pioneered the financing of pilot schemes, amongst others, for integrated rural development, notably in Eastern Region, and also major training programmes for staff in the major sectoral ministries, such as Education, Health, Agriculture, Transport and in the Ministry of Planning itself, responsible for implementation of the decentralization programme. In particular, the Ministry of Planning has been responsible for the selection and training of District Secretaries, the key figures as Chief Executive and Member of the District Assemblies, and for the formulation of the legal bases of control over, and responsibilities of, the districts. The training programmes have involved headquarters staff in these ministries, but much more importantly they have been targeted at the considerable training needs of field staff in a decentralized system.

This is not the place to begin even a general evaluation of the decentralization programme. It would be much too early to do so, given that District Assemblies are only recently elected and will take some time to establish their role in development, *vis-à-vis* national government and also traditional power structures. Much will depend on practice and how the programme evolves, rather than on statutory powers and responsibilities. The turn out of eligible voters in the non-political district elections was under 60 per cent, hardly an overwhelming enthusiasm. The majority of those elected in Ashanti, Eastern, Central and Western Regions were farmers or teachers: 28 farmers and 13 teachers out of 47 elected in Sekyere District, Ashanti; 18 farmers and 15 teachers out of 38 elected in Assum District, Central Region (*West Africa*, 19 December 1988). Members nominated by government are mostly public officials from a wide variety of technical and functional backgrounds. Speeches to the first meetings of the assemblies by PNDC members, including the President, encouraged members to focus on 'youth, rural employment and resource generation for development' (*West Africa*, 30 January 1989).

Decentralization and educational planning

Ghana has been at the forefront of educational development in Africa since long before independence and, by the 1950s, it had probably the highest proportion in Africa of its children in school, with a solid qualitative base and rapidly expanding enrolments at secondary and higher levels. The rate of expansion increased sharply with independence, as Nkrumah's vision of an industrialized Ghana leading the Pan-African movement towards continental self-sufficiency recognized the need to have that leadership underpinned by a well-developed human-resource base. Expansion was guided by successive manpower plans, based on what have turned out to be highly optimistic assumptions of the demand for skilled labour, that were formulated within the highly centralized government planning structures, a situation that found a parallel in most African countries at that time (see Jolly, 1969; Jolly and Colclough, 1972). With continuous economic decline to 1983, however, the supply of educated manpower seemed to outstrip demand greatly, and not only were there major problems of low wages and unemployment within Ghana but also a massive and continuing exodus of skilled manpower to other African countries, particularly Nigeria, and classic 'brain-drain' migrations to Europe and North America (Gould, 1985).

Between the mid-1960s and Phase II of the ERP, 1987–9, the education system in Ghana had undergone little structural change. Its planning base was weak for the system was driven largely by the level of demand for education. The level of individual demand for education had been inflated by the expectations of returns on investments in education created in the early years of independence. The Ministry of Education itself was traditionally weak and remained so after independence. In the colonial period, the various Christian missions rather than the government were the leaders in education. After independence the government became more directly involved but through semi-autonomous implementing agencies: the Ghana Education Service (responsible for primary and secondary schools) and the Higher Education Division. All educational planning, mostly incremental and short term, was highly centralized within these sections at ministry headquarters in Accra. Plans for structural reform to strengthen the central policy role of a co-ordinated ministry did exist, but governments had neither the political will nor the necessary finance to implement them. As the national economy deteriorated, so did the state of the schools. As late as 1976, education consumed 6.5 per cent of GDP, but that figure had fallen to 1 per cent of a reduced GDP in 1983, though it rose to 1.7 per cent in 1985. In 1976, non-salary costs were 19 per cent of the recurrent budget in education, but these had fallen to 6 per cent by 1986. Enrolments declined slightly over this period (despite population growth at *c.* 3 per cent per annum), but quality of education fell dramatically, starved of financial and learning resources and with major emigration of teachers, especially from universities and secondary schools.

Clearly the ERP needed to halt the decline of education. It introduced changes that sought a more positive contribution by schools to the economy and society of Ghana. In the last five years, major structural changes have been effected in four areas. In the first place, the overall responsibilities within government have been reorganized so that there is now a single Ministry of Education and Culture. The integration of culture into education is indicative of the political desire within PNDC for a reorientation of the role of education and of culture in national development objectives. Second, the overall shape of the formal school system has been restructured and its length reduced from what for many was a 17-year schooling cycle (8–4–5), one of the longest in the world, to 12 years (6–3–3), with particular attention being paid in the earliest phase of change to the organization and curriculum of the three-year junior–secondary cycle. A third major change has been the strengthening of the ministry, as a central co-ordinating and planning agency, relative to the implementation agencies. The creation of a strong Planning, Budgeting, Monitoring and Evaluation Division (PBME) under the direct control of the Secretary for Education and Culture (i.e. the minister and political head) provides the means of formulating policy and assessing and controlling financial allocations and structures to implement policy and to monitor progress. Fourth have come changes associated with decentralization, with the District Assemblies assuming responsibilities for many aspects of basic education (primary and secondary) including the financing and day-to-day running of schools, the posting of teachers, maintenance of buildings, the formulation of annual expenditure estimates and the preparation of annual and medium-term District Education Plans.

These changes are all consistent with, indeed will provide a testing bed for, many of the educational policies of the Human Resources Division of the World Bank, as elaborated in its policy study, *Education in Sub-Saharan Africa: Policies for Adjustment, Revitalization and Expansion* (1988), that itself must be seen as a necessary component of structural adjustment in Africa as a whole. Since major investments in education are economically sound, indeed returns to both primary and secondary education seem to yield higher returns than investments in most other sectors (Psacharopoulos and Woodall, 1985), much increased investments in Ghana's underfunded system are easily justified. But it should not simply be more of the same. There needs to be more control on expenditure for educational costs are high as a proportion of government expenditure. Since structural adjustment policies seek to reduce the direct expenditure of the state, there is much discussion of generating alternative sources of finance and reducing unit costs, at least those costs borne by the government. In Ghana this has meant 'cost sharing', i.e. passing some of the costs from the public purse to the individual consumer. This has been pioneered in the 'deboardingization' programme for secondary schools (converting high cost boarding schools into lower cost day schools, and passing the costs of accommodation from the school to the pupil), in the sharp reductions of student grants and introduction of loans in the tertiary sector, both of which inevitably favour the rich rather than the poor student. Parallel moves in the health sector include the introduction of sharply increased user-charges for health service, with serious adverse effects on equity of access to them (Waddington and Enyimayew, 1989). Yet, at the same time, there is concern to expand primary school enrolments (currently standing at only 70 per cent of the appropriate school-age cohorts) especially for girls, and to restructure the curriculum to emphasize practical and life skills rather than academic subjects, all of which are dependent on significant inputs of foreign assistance. The World Bank will lend US \$25 million for the period 1987–92 to the education sector in Ghana.

For the purpose of this chapter, however, it is appropriate to pay particular attention to the third and fourth objectives of education reform in Ghana: the central planning responsibilities of the ministry and the responsibilities of the districts, and the relationships between those two levels. Is decentralization in education to 110 districts possible or realistic where there is an enhanced central planning capability in the ministry in PBME Division? As in all social and economic sectors, the ERP is allocating resources to enhance performance and capability through training of planning officials and through physical provision (vehicles and office facilities including computers) at both levels, on the assumption that there is a complementarity between the centre and the districts. Stronger central control is needed for strategic planning and setting pedagogical and financial norms and objectives, but local responsibilities must lie in day-to-day allocation decisions based on systematic analysis of local needs and requirements, set within the national planning norms and other criteria. Central government can also monitor local performance and offer specialist assistance and training where required. Where local responsibilities are associated with local needs, the needs can be met within the limits of available finance and other necessary inputs. Assessment of local needs in education will then be based on more realistic criteria and a better formulated sense of priorities, with more pressure to contribute to cost sharing, using the revenue-raising powers of the District Assemblies as well as direct levies from consumers of the service.

In addition, however, the Ministry of Education may cede some direct control over schools and expenditures on education to the District Assemblies operating at an interministerial local level rather than to district education staff. In a country such as Ghana, which has a tradition of discrete sectoral organization of government activity inherited and largely unaltered from the colonial era, with relatively little interministerial co-operation or co-ordination except at the highest level, the loss of sectoral authority is keenly felt. Composite budgeting by districts is seen somehow as a threat to the strength of the education sector – by far the largest and best organized sector at the district level. In practice, however, composite budgeting may offer an opportunity for more rather than fewer resources to education through the decisions of the assemblies.

The fundamental shift is therefore from a centralized *sectoral* planning base, in which all planning decisions and expenditure priorities were justified in terms of criteria established centrally within the Ministry of Education, to one in which there is *spatial* planning. In the new structure there are inputs from central ministries on technical criteria, with flexible guidelines as well as formal requirements in some aspects (e.g. curriculum), but with decisions on intersectoral as well as some intrasectoral expenditures being made by the District Assemblies on the basis of technical advice from local planning officers. Throughout Africa there are tensions between the more traditional sectoral and more innovative spatial strategies in educational planning and these have been resolved in a range of ways, generally consistent with wider regional development strategies operating in a particular country (Gould, 1988a). However, these tensions are most evident in three areas: interdistrict inequalities in quantitative indices, the issue of interdistrict variation in curriculum and other qualitative issues and the planning powers of the districts *vis-à-vis* the centre.

Spatial variations in educational provision and enrolment at the regional scale were apparent even after the massive expansions of the 1950s and early 1960s (Hunter, 1964) and have remained up till the present time (Pandit and Asiamah, 1988). Largely as a result of problems of disaggregating data to the district scale (neither population nor enrolment data are yet available for the newly created districts), it is not possible to consider spatial variations in education at the district scale, but it would be surprising indeed if intraregional (i.e. interdistrict) inequalities were not greater than interregional inequalities. Within each region in Ghana, as experience elsewhere has shown (Carron and Chau, 1980), there are known variations in school provision and attainment, with higher levels in and near the regional capitals and towns generally than at the rural periphery. This situation may not be ameliorated after decentralization for two rather different reasons. First, those districts that are educationally less well developed are also likely to be poorer and be less able to add incremental expenditures to the base-line provision either from locally generated public revenues or from private incomes. Second, poorer areas are less likely than richer areas to see education as a major priority, preferring directly economic activities to boost income levels in the short term.

One major area of activity of PBME Division must therefore be to monitor the extent of variation in enrolment and other quantitative indices (e.g. pupil:teacher ratio, drop-out, promotion and repetition rates). It must be able to exercise control over aggregate allocations and to establish norms (e.g. upper and lower limits for enrolment rates at any given level) within which districts should fall. Unless there is a centrally co-ordinated perspective on interdistrict variation, and there are the means to remedy, if not eliminate, inequalities, then it is certain that the extent of disparity will grow in the immediate future: the richer areas will achieve and will themselves be able to purchase increasingly preferential resource allocations.

The problems are similar in many respects for qualitative indices of educational status, especially in so far as they relate to non-salary recurrent expenditures such as textbooks, furniture and learning materials generally. Over 95 per cent of recurrent expenditure in schools comprises teachers' salaries. There is a severe shortage of learning materials in Ghanaian schools and this has been a major factor in falling quality. However, expenditures vary enormously from school to school, even within small areas, and are often directly related to individual benefactions or money-earning activities of an enthusiastic parents' committee. Hence the need for monitoring the qualitative as well as quantitative indices of provision between districts.

In other important respects, however, decentralized decision-making over aspects of the curriculum that can give rise to interdistrict variations can be a source of strength worth encouraging. In a country such as Ghana, with a multiplicity of local languages and cultures, a strictly national and centralized system imposed a rigidity that may not be appropriate to local circumstances. Though English is the medium of instruction at most levels, early instruction has been given in a small range of the larger Ghanaian languages. These have become dominant at the expense of local languages and, by implication, of cultures that may be localized within a district, but form a minority within a region. District educational planning may be more able to encourage and foster local languages, together with dancing and other aspects of local culture. In the economic sphere, vocational and practical subjects may be chosen to suit the local economy, and even the school year can be adjusted to local demands for children's labour at peak times of harvest that can be variable over small areas. As Pierre Furter (1980) has argued for his native Switzerland, a decentralized education system that has been sufficiently flexible to take account of and encourage local distinctiveness has been a major factor in the integration of the Swiss state. Rather than attempting to impose a national culture and inflexible school procedures centrally, trends that were likely to be dominant in the period soon after independence when nation building was an important objective of the education system, the time may now be right in Ghana to reassert subnational cultures. The integration of a Ministry of Education and Culture has made the need for cultural awareness in educational planning quite explicit, and decentralization of some decision-making may be a necessary component of that distinctiveness.

The third issue of specific concern in this context is the relationship between planning responsibilities at the centre and in the districts. These have still to be defined precisely, but local responsibilities will include decisions over the number, size and distribution of schools. In a highly centralized system, as in the past in Ghana, there is a uniform model school and a class size. Norms can be established centrally, and implemented by local officials who have little scope to make decisions or exercise their own initiative, being required to follow the centrally imposed criteria slavishly. In practice, however, in a school system the size of Ghana's, and with relatively weak systems of control, very wide variations in school and class size exist within and between districts, including a large level of 'inefficiency', as measured by high costs per pupil. Given the very high proportion of costs consumed by teachers' salaries, maintaining a high pupil:teacher ratio and relatively large classes is essential if costs are to be minimized. The official 'norm' for maximum primary school class size in Ghana is 46, yet the national mean pupil:teacher ratio in primary schools in Ghana is 23:1 with a range from 18:1 in Northern Region to 31:1 in Greater Accra (Table 12.1, column 5). To some extent the lowest ratios are attributable to low population densities, e.g. in Northern Region, so that schools and classes may need to be considerably smaller than the norm where there are insufficient pupils for the norm within the catchment of a school (Gould, 1982). However, the main reason for low pupil:teacher ratios and small schools in much of rural Ghana is that, due to mission and denominational rivalries, there are too many schools, with overlapping catchment areas. Centralized planning has proved unable to grasp the nettle of denominational duplication. The church or mosque groups and their schools' managers dictate the allocation of teachers even though they are paid by the government. Where teacher allocation is devolved to district authorities and local communities are allocated teachers on the basis of a feasible and locally enforceable norm, then communities will need to maintain high pupil:teacher ratios or, alternatively, to provide additional resources themselves to pay the extra teachers necessitated by duplication and smaller classes. This, it is hoped, will provide an incentive that has not previously existed for co-operation at the local level to reduce unit costs. Thus decentralization of many allocation decisions, such as who will need to pay the extra costs of 'inefficient' allocations, will further the World

Bank goal of unit cost reduction without further loss of quality. Indeed, since half the teachers in primary schools in 1988 were untrained, more efficient use of the trained teaching force may improve the quality of classroom teaching. Local diagnosis of the pattern and functioning of schools may transform the local geography of education provision.

Conclusion

Since the decentralization programme is still being developed in detail, since District Assemblies have only recently been elected and since their priorities are not yet established, it is clearly much too early to make any realistic assessment of policies as they affect education or other sectors in the wider context of structural adjustment and the ERP. In which respects and to what extent will decentralization complement other measures in the economic and social spheres? More information is needed not only on the districts themselves (population, economic activity, social provision, etc.) but also on the experience of each district in establishing a purposeful framework for harnessing local resources, of labour and land as well as capital. Geographers, amongst others, will be concerned to monitor progress in a range of districts in the next few years as the local and micro-impacts of macrostructural adjustment policies in general are evaluated.

With particular respect to the education sector, however, it is clear that there are major advantages in decentralizing decision-making in educational planning to the local level. Local control of some aspects of education – its finance, buildings, the relationship between schools and the communities they seek to serve – can contribute both to quantitative expansion necessary if better access to schools, especially for the poor, is to be achieved, and also to qualitative improvement, through greater availability of teaching materials and more specific concern for the contribution schoolchildren and schoolleavers can make towards the needs of the community. Yet in a country like Ghana, where the education sector is large and strong and traditionally highly centralized with a long standing tradition of top down decision-making, the establishment of a system dependent on a complementarity between bottom up and top down planning responsibilities will inevitably lead to tensions and conflicts. The revitalized Ministry of Education and Culture has certainly taken a positive approach to decentralization, with a major nationwide training programme for the newly appointed district education staff with planning responsibilities. These include the District Assistant Directors (notionally in charge of the primary- and secondary-schools system in each district) as well as their Statistics Officers. The training has identified both major shortcomings in the current planning capabilities and experience of these officials, but also and more seriously, it has revealed major information gaps about what their new powers and responsibilities are to be and how these will be exercised. Officials who in the past have had few planning responsibilities and who, in the area of statistics, for example, have merely transmitted school data collected by standard school censuses upwards to regional and national headquarters, now find that they are responsible for making diagnoses and allocative decisions based on these statistics, regarding teacher postings or the distribution of new textbooks, for example. The PBME can offer technical support and broad guidelines, but must be prepared to permit greater diversity than hitherto; the districts are able and prepared to make technically justifiable decisions based on PBME criteria. If the partnership between the PBME Division of the ministry and the districts can be recognized on both sides then the new decentralized structures will be able to raise the efficiency of, and improve equality of access to, the education system.

Over thirty African countries are currently implementing structural adjustment measures, usually with the support of the World Bank and the IMF (UNECA, 1989, p. i). The majority of them have experienced adjustment in the education sector, and most, implicitly or explicitly, are introducing more decentralized responsibilities in the public sector. All recognize that macro-adjustments have major internal effects both socially and spatially, and in Ghana these spatial effects are bound up with the decentralization programme. Success in implementing a decentralized education system in Ghana would show that there can be structural adjustment with a human face, making a major contribution to social improvement. Success in Ghana, as elsewhere, will certainly need continuing large resource inputs into the schools in the short term to raise quality and expand access. These resources will need to come from external sources if there is not to be increasing differentiation between rich and poor within each district and between richer and poorer districts. The speed with which resources are switched from external support to internal 'cost recovery' may be crucial here. If this is too rapid, without the macro-economic policies having had their desired effect on supply-side factors of rural development and income generation, then the decentralized education system will be, even more than it has been in the past, a major symptom of spatial differences and a prime vehicle for perpetuation of social differentiation. The benefits of increased local production, both in the aggregate and locally, anticipated by the ERP but not yet apparent, must be channelled into social sectors that continue to spread their impact beyond the direct beneficiaries of increased production. This may mean an increase in spatial diversity as a result of the decentralized decision-making. It will add further complexity to the geography of Ghana, but it will be well worth while.

References

Allison, C. and Green, R. H. (eds.) (1983) Accelerated development in Sub-Saharan Africa: what agendas for action?, *IDS Bulletin*, Vol. 14, no. 1.

Aryeetey-Attoh, S. and Chatterjee, L. (1988) Regional inequalities in Ghana: assessment and policy issues, *Tijdschrift voor Economische en Sociale Geografie*, Vol. 79, no. 1, pp. 31–8.

Carron, G. and Chau, T. N. (eds.) (1980) *Regional Disparities in Educational Development*, UNESCO, International Institute for Educational Planning, Paris.

Colclough, C. and Green, R. H. (1988) Do stabilization policies stabilize?, *IDS Bulletin*, Vol. 19, no. 1, pp. 1–5.

Cornia, G., Jolly, R. and Stewart, F. (eds.) (1987) *Adjustment with a Human Face: Protecting the Vulnerable and Promoting Growth*, Oxford University Press.

Ewusi, K. (1987) *Structural Adjustment and Stabilization Policies in Developing Countries: A Case Study of Ghana's Experience in 1983–1986*, Ghana Publishing Corporation, Accra.

Furter, P. (1980) The recent development of education: regional diversity or reduction of inequalities?, in G. Carron and T. N. Chau (eds.), op. cit., pp. 49–113.

Ghana Government (1987) *National Programme for Economic Development (Revised)*, Accra, 1 July.

Ghana Government (1988) *The PNDC Budget Statement and Economic Policy for 1988*, Accra, 16 January.

Gould, W. T. S. (1982) Provision of primary schools and population redistribution, in J. I. Clarke and L. A. Kosinski (eds.) *Redistribution of Population in Africa*, Heinemann, London, pp. 44–9.

Gould, W. T. S. (1985) International migration of skilled labour within Africa: a bibliographical review, *International Migration*, Vol. 23, no. 1, pp. 5–28.

Gould, W. T. S. (1988a) Regional planning and educational planning in Sub-Saharan Africa, *Regional Development Dialogue*, Vol. 9, pp. 39–57.

Gould, W. T. S. (1988b) Rural–urban interaction and rural transformation in Tropical Africa, in D. Rimmer (ed.) *Rural Transformation in Tropical Africa*, Belhaven, London, pp. 77–97.

Green, R. H. (1988) Ghana: problems and limitations of a success story, *IDS Bulletin*, Vol. 19, no. 1, pp. 7–15.

Hough, J. R. and Bossman-Dampare, K. (1988) Budgetary analysis using education and expenditure ratios (unpublished MS, Planning, Budgetary, Monitoring and Evaluation Division, Ministry of Education and Culture, Accra).

Hunter, J. M. (1964) Geography and development planning in elementary education in Ghana, *Bulletin of the Ghana Geographical Association*, Vol. 9, no. 1, pp. 55–64.

Jamal, V. and Weeks, J. (1988) The vanishing rural–urban gap in Sub-Saharan Africa, *International Labour Review*, Vol. 127, no. 3, pp. 271–92.

Jolly, R. (1969) *Planning Education for African Development*, East African Publishing House, Nairobi.

Jolly, R. and Colclough, C. (1972) African manpower plans: an evaluation, *International Labour Review*, Vol. 106, no. 2, pp. 207–64.

Jonah, K. (1987) *The Social Impact of Ghana's Adjustment Programme, 1983–86*, IFAA, London.

Kasanga, R. K. and Avis, M. R. (1988) Internal migration and urbanization in developing countries. Findings from a study of Ghana (research paper no. 1, Department of Land Management and Development, University of Reading).

Moseley, P. and Smith, L. (1989) Structural adjustment and agricultural performance in Sub-Saharan Africa, 1980–87, *Journal of International Development*, Vol. 1, no. 3, pp. 321–55.

Pandit, H. N. and Asiamah, F. (1988) Forecasting enrolment and teaching manpower demand for basic education in Ghana (1987–2000) (unpublished MS, Planning, Budget Monitoring and Evaluation Division, Ministry of Education and Culture, Accra).

Potter, R. B. and Unwin, T. (eds.) (1989) *The Geography of Urban–Rural Interaction in Developing Countries*, Routledge, London.

Psacharopoulos, G. and Woodall, M. (1985) *Education for Development. An Analysis of Investment Choices*, Oxford University Press for the World Bank, Washington, DC.

Stevens, C. and Killick, T. (1989) Structural adjustment and Lomé IV, *Trocaire Development Review*, 1989, pp. 25-40.

Taaffe, E. J., Morrill, R. and Gould, P. R. (1963) Transport expansion in underdeveloped countries: a comparative analysis, *Geographical Review*, Vol. 53, no. 4, pp. 503–29.

Todaro, M. P. (1976) *Internal Migration in Developing Countries*, ILO, Geneva.

UN Development Program (1987) *Fourth Country Program for Ghana*, New York, NY.

UN Economic Commission for Africa (1989) *African Alternative Framework*, Addis Ababa.

Waddington, C. J. and Enyimayew, K. A. (1989) A price to pay: the impact of user charges in Ashanti-Akim District, Ghana, *International Journal of Health Planning and Management*, Vol. 4, no. 1, pp. 17–47.

World Bank (1981) *Accelerated Development in Sub-Saharan Africa: An Agenda for*

Action (the Berg Report), Washington, DC.

World Bank (1984a) *Ghana: Policies and Program for Adjustment*, Washington, DC.

World Bank (1984b) *Towards Sustained Development in Sub-Saharan Africa: A Joint Program of Action*, Washington, DC.

World Bank (1988) *Education in Sub-Saharan Africa: Policies for Adjustment, Revitalization and Expansion*, Washington, DC.

World Bank (1989) *Sub-Saharan Africa: From Crisis to Sustainable Growth*, Washington, DC.

13

DATA NEEDS AND DATA SYSTEMS FOR REGIONAL PLANNING IN DEVELOPING COUNTRIES: A COMPARISON OF EXPERIENCES IN KENYA AND SRI LANKA[1]

Jan J. Sterkenburg

Introduction

There is a clear tendency towards decentralization of development planning in many developing countries. Expected advantages of this decentralization are the improvement of efficiency in the planning organization, by providing a means of co-ordinating the various agencies involved, a greater flexibility in plan implementation, increased participation by the local population and the better adjustment of policy measures and development activities to local circumstances (Conyers, 1983; Conyers and Hills, 1984; Hyden, 1985). The latter factor in particular is based on the planners' more detailed knowledge about the conditions, needs and development options in their area, especially about the availability of reliable information in accessible form for the various organizations involved in planning and implementation of measures instrumental to regional development (see also Sundaraman, 1987).

The objective of this chapter is to assess the data needs, to describe the characteristics of possible database systems for regional development planning and to identify the major problems in establishing a database. It attempts to realize this by describing and analysing the situation and experiences in two regions with rather different conditions, namely, Kakamega District in Kenya and Nuwara Eliya District in Sri Lanka. However, before dealing with the differences between the two regions, some similarities in the environment for planning for the two countries should briefly be mentioned.

Both Kenya and Sri Lanka were seriously hit by the global recession of the late 1970s to early 1980s, which led to a reduction of government expenditure, an increase in dependence on foreign aid and a succumbing to the demands accompanying structural adjustment loans. In addition, there is a remarkable similarity in the type of economy and in the nature of development policies: the economy of both countries is dominated by the agricultural sector, which shows a dual pattern of large estates and smallholders, agricultural exports being an important foreign-exchange earner. Moreover, both countries are pursuing growth oriented policies but with different degrees of state intervention.

The intensity of state intervention is related to the nature of the state that, among other things, is evident from the legal and institutional position of political parties, including the acceptance of political opposition. In Kenya, the multiparty state with a relatively high degree of decentralization was turned into a one-party state within a few years of independence. With a strongly dominant executive presidency, Kenya showed a high degree of political stability until an abortive army coup in 1982. Since then the single party's political control has increased and the administrative structure has become more centralized. Sri Lanka is a multiparty state and regular elections determine the relative strength of each party in parliament. The two main parties are dominated by representatives of the ruling élite, a limited number of influential families. The opposition against this élite – representing the poorer sections of the rural population – is mounting and becoming increasingly violent. This class conflict is further complicated by the ethnic strife between the Singhala and Tamil populations.

In both Kenya and Sri Lanka, decentralization of development planning is influenced by a sheer necessity of steering the development process more successfully, by political motives in relation to ethnic differentiation and by donor pressure. Yet, both Kenya and Sri Lanka demonstrate resistance to a genuine transfer of power to lower echelons of administration and elected bodies. This is reflected in particular by the limited financial scope for district planning organizations and the strong position of sectoral agencies at the district level.

Factors influencing a database for regional planning

Regional planning is understood here as a decentralized and integrated type of development planning that comprises elements of both economic planning and physical planning. The decentralization aspect refers to the transfer of authority to subnational levels, the geographical units usually being administrative ones (provinces or districts). The integration in the planning process refers to the combination of economic and spatial aspects of planning and to their relationship to various sectors of the economy. This type of planning requires a reliable database at the regional level in order to produce relevant information for planning purposes. A distinction is made between data and information. Data comprises a registration of facts, whereas information refers to a specific use of processed data for planning purposes, and ultimately for a planning decision.

No standard formula can be given for the type of data to be incorporated into the database and their detail and magnitude, although certain types of information will be required for any planning effort. At the general level one may think of basic data on natural and human resources, the organization of production, the employment and income situation and the quality and accessibility of productive and community services. Yet, such broad categories cannot function as a proper yardstick for a database: they merely give a first indication of the type of data that may be included. A further specification of the data needs for a given region first requires more information about factors influencing the nature and the intensity of the planning process at the regional level. Three factors may be singled out as being of particular importance in this respect, namely, government policy, the fundamental characteristics of the region and the quality of the planning organization.

At least two aspects of *government policy* influence the need for data at the regional level, namely, the type of development policy apparent from the intensity and nature of government intervention and the degree and type of decentralization. The former determines the scope for intervention on the part of the planners and government institutions. The latter indicates to what extent there is a genuine transfer of authority and power to lower levels of administration and elected bodies, and especially how and to what degree the financial means become available for the implementation of concrete activities. Both elements appear most clearly in the nature of the planning process and the type of plan document produced. By definition, process planning works with less detailed data at the beginning of the intervention and expands the database by experience obtained in executing activities, whereas blueprint planning makes use of a more exhaustive database. In addition, the type of plan influences the nature of the database: skeleton and inception plans are less demanding than comprehensive district plans.

The *type of region* as a factor influencing the nature and magnitude of a database should be understood mainly in terms of the region's complexity as appearing from the variety in natural resources, population density and distribution, the differentiation in economic activities and social groups and the degree of urbanization.

The quality of the *planning organization* refers to the composition of the planning unit as indicated by the types of expertise and the variety of disciplines, the level of training and the degree of experience and skills and, finally, again, the authority transferred to this organization to undertake the co-ordination of sectoral activities at the regional level.

In sum, a standard list of required data for a decentralized or district level database cannot be produced, except for certain broad categories of information needed for any type of planning. However, the amount of data should be kept to a minimum. The challenge is to facilitate effective planning with a minimum of crucial data, and without burdening the planning officers with a Herculean task in data gathering, processing and storage. The specific data needs are determined by local conditions as indicated by the nature of government policy, the type of region and the characteristics of the planning organization. These three factors will be given attention in some more detail for two regions: the Kakamega District in Kenya and the Nuwara Eliya District in Sri Lanka. In addition, the nature of the databases established for regional planning purposes is described.

Decentralized development planning in Kenya

Government policy

Kenya's development policy can be typified briefly as growth oriented with a relatively high degree of government intervention in production and marketing, in prices and wage levels and in the allocation of resources. The policy has a pronounced urban orientation, which is evident from the declining share of agriculture in investments up to the middle of the 1970s, the priority to the urban areas in the supply of social and infrastructural facilities and the higher real wages and incomes in the urban areas. In agriculture, the emphasis has been placed on the support of the large farms and export crops. Implicit in this policy is the assumption of a trickle down of the benefits of growth to all groups in society in the long run. The evidence to the contrary – as produced in the 1972 ILO report and other studies – has resulted in a rephrasing of policy documents but has also brought little change to the actual situation (ILO, 1972; Livingstone, 1986). Therefore, socioeconomic inequality increased strongly during the post-independence period, a phenomenon closely related to the uneven distribution of productive resources, land in particular (House and Killick, 1983). The redistribution of part of the white settlers' land and the incorporation of marginal areas into an Arid and Semi-Arid Lands (ASAL) Programme has not prevented an increase in landlessness and poverty in the rural areas. Present policy emphasizes intensification of land use by small scale farmers with priority for the traditional export crops, coffee and tea, and support to the informal sector, but it evades a discussion about the need for a more drastic land reform. The present decentralization policy is an instrument to facilitate more detailed planning for the intensification of agriculture and for support to informal sector growth in the form of Rural Trade and Production Centres (Republic of Kenya, 1986).

Decentralization policy in Kenya is a rather recent phenomenon. The first decade after independence was characterized by centralization and by a shift of power to central government institutions in the national capital. From the mid-1970s onwards a gradual transfer of authority to the lower echelons of government administration may be observed. This appears from the appointment of District Development Officers (DDOs), the establishment of District Development Committees (DDCs) as the bodies to set priorities at the district level, the introduction of District Development Plans for five-year periods according to standard guidelines and a number of cautious experiments with integrated rural and regional development planning supported by donor agencies – chiefly ASAL programmes in the marginal areas. (For a detailed account of the history of Kenya's decentralization policy, see Cohen and Hook, 1986.)

In 1983 a further specification of Kenya's decentralization policy was announced under the name District Focus Policy for Rural Development. The essence of the strategy is a further transfer of power to district-level government staff. The co-ordinator of the strategy at the district level is the District Commissioner, a political appointee, technically supported by the DDO who, as the head of the District Planning Unit (DPU), is also the secretary of the DDC. The committee is dominated numerically by the heads of the departments, i.e. by government-appointed technical officers. The committee plays a crucial role in formulating the five-year district development plans and the related annual investment plans. It is assisted by a smaller technical committee, the District Executive Committee. At lower levels of administration, i.e. the division and the location, divisional and locational development committees have been established.

From 1986 onwards the planning process at the district level has been supported by District Information and Documentation Centres (DIDCs). The DIDC is seen as an important component of the District Planning Unit, which provides essential information to the DDC, other development committees and the heads of departments. It is considered a major source of information on development statistics, national and district plans, research reports and other types of technical information for each sector. Circulars about the type of documents to be included and about the training programme for DIDC staff reveal that the DIDC is seen at present chiefly as a reference library for district staff involved in development planning.

Kenya receives donor support for its decentralization policy. Advisers to the Ministry of Planning and National Development are made available through USAID, whereas The Netherlands and Scandinavian countries assist in the implementation of the ASAL programme. In addition, The Netherlands supports the Kakamega District Focus Project financially.

Kakamega District Focus Project

The Kakamega District Focus Project aims to support the decentralized development-planning process in Kenya by assisting the Kakamega DPU in establishing a DIDC, setting up a database for district planning and identifying bottlenecks in the development process in the area. The experiences in the district are directed towards the design of a methodology replicable in other parts of the country. In more concrete terms, the project comprises the construction of a DPU/DIDC building, the provision of equipment for the building, the training of district planning staff and the collection of documents and data-facilitating development planning in the Kakamega District.

Kakamega District, which is part of the Western Province, has an area of 3,250 km². It consists of a hilly zone intersected with deep river valleys in the south at an altitude of 2,000 m, and a slightly undulating peneplain in the northern, eastern and central parts with an altitude varying between 1,100 and 1,900 m. The district has a high annual rainfall, with 1,500–1,800 mm in the south, gradually decreasing to 1,000 mm in the north. The rainfall is well distributed over the year, especially in the south, but the northern-most part experiences a longer dry season. In general

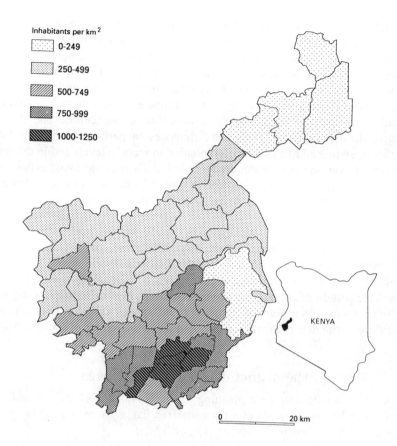

Figure 13.1 Population density per location, Kakamega District, 1988 (*Source:* DIDC/GIS, 1987)

rainfall, far more than soil quality, enables the district to qualify as a high-potential area. This partly explains the high population density. With a total population of over 1.4 million and a density exceeding 400 persons per km² on average, the district is one of the most densely populated parts of the country. However, densities differ greatly between the various parts of the district, and generally decrease from south to north (Figure 13.1). Differences in soils, climate and population density are associated with variations in farm size, cropping patterns and the role of livestock. Maize is the dominant foodcrop in all parts of the district. Maize surpluses in the northern part are transported to the south and central-west to make up for the deficits there. These deficits result from the landlessness and small farm sizes and from the cultivation of cash crops, such as coffee, tea and French beans in the south, and from the cultivation of sugar cane in the central and western parts. Apart from maize, dairy farming is an important source of income in the north. The district has little non-agricultural employment, a factor that also explains the low rate of urbanization. In view of the high population density and the absence of adequate employment opportunities in the district, both in agriculture and in non-agricultural activities, the Kakamega District is known for a high rate of migration chiefly circular in nature. An estimated one third of the active population finds permanent and temporary employment outside the district. Nairobi, Mombasa and other urban areas in Kenya are the main destinations, but migrants also go increasingly to rural areas with a high rate of commercialized agriculture and large farms to find employment and augment their incomes. The spatial variation in farm size, type of agriculture and degree of commercialization, together with the differences in participation in the labour-migration system lead to a sharp differentiation in income levels and in the relative importance of various income sources for individual households. And as households show a high degree of adaptability to changing circumstances in accordance with the availability of labour in the household, the situation may alter quite rapidly over time.

This complex and dynamic situation already puts a high demand on the information needs for planning and the related database, and also on the skills of the planners to design realistic plans for the district's development. This demand for reliable data increases further in terms both of magnitude and complexity if the intricate pattern of community services (schools, health facilities, water-supply systems) is taken into account, together with the spatial variation in quality of the physical infrastructure required for the adequate functioning of these services.

The district database in Kakamega

The database for development planning in the Kakamega DIDC, in addition to the reference library, consists of a computerized data system. This data system is composed of the following elements (Figure 13.2):

- *General data on population* Demographic data originating from the 1979 population census and supplemented by estimates and projections for later years, for all locations in the district.
- *Data on natural resources* Data on soils, climate and land qualities available for a variety of geographical units, supplemented with soil suitability data for main crops for selected locations.
- *Location specific data (LSD) formats* A comprehensive set of data for all locations in the district on land use, crop acreages, livestock, agricultural services, physical infrastructure and community services.
- *Sectoral production data* Key data on main cash crops in the agricultural sector, such as area under crop, number of growers and total output for all relevant locations.
- *Physical infrastructure data* Data originating from a physical-infrastructure inventory preceding the design of the 1989–93 district development plans, including the main characteristics of roads, cattle dips, co-operative societies and market centres, grouped per location.
- *Community services data* Data referring to health and education services, initially collected as part of the physical-infrastructure inventory and sub-sequently expanded and incorporated into LSD formats.

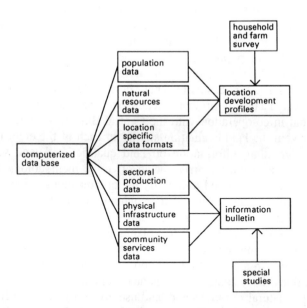

Figure 13.2   Database for Kakamega District

The construction of the database is undertaken using a variety of methods and with reference to different geographical units. General data on population and resources are obtained chiefly from secondary sources, such as the population census, meteorological reports and soil survey reports and maps. The sectoral database on agriculture requires field visits to the organizations involved in input supply and marketing of specific crops, such as the coffee co-operatives and the tea leaf collection centres. LSD formats consist of a set of sheets to be completed by government officers stationed at the administrative levels of the division, location and sublocation. A major principle is that the officers are not assigned with extra tasks. Most of the data are already collected in the course of their current reporting routines, and the formats and the spatial units are standardized for all departments. In addition, the full set of sheets is only administered to establish the database, and again in the year preceding the new district development plan with updates for selected items. This enables the district planners to base their plan on recent data and to indicate trends over the preceding period.

The geographical units used in the database are those coinciding with the administrative areas of the district, the division and the location. These units come into a single hierarchy, and they have also been selected as units in the development-planning process. At the division and location levels, development committees have been set up as well, although little is known about their functioning as yet. Irrespective of that, the need for their participation in data collection – as a first step in the planning process – is obvious. The data system is being tested and supplemented by means of a series of special studies. Pilot surveys on households and farms have been carried out in eight locations covering the variety of ecological zones and spatial differences in population density over the district. The results of these surveys supplement the LSD formats from the perspective of the household and the farm. For example, the LSD formats give the number of schools, the number of pupils by sex and the number of teachers by level of training. The household survey indicates the degree of school attendance of children in the school-going age by household. These household and farm surveys are also undertaken to design a useful format replicable over the whole district successfully, to be carried out in co-operation with the Central Bureau of Statistics.

In a similar vein the land evaluation studies in each of the eight locations not only focus on providing information on land qualities and land suitability for selected crops but also aim to arrive at an effective methodology for land use planning at the district level and to identify the organizations at the national level able to support this element of the district planning exercise.

The data collected and stored in the database are processed and analysed at present in the form of two types of planning documents, namely Location Development Profiles and special studies. Location Development Profiles are brief documents that contain a general description and basic information and maps for individual locations. They combine the updated estimates of the population census, the results of the natural resources and land use inventory, the data from the LSD formats and those of the household-cum-farm survey. The document concludes with a summary of the location's major development problems. The Location Development Profiles may be grouped together into Division Development Profiles and in this way become important building-blocks for a district development plan.

Special studies deal with specific aspects or issues at the district level. Examples of such studies are a soil suitability analysis for major crops, studies on the livestock and health sectors, a study on landlessness and land concentration, on migration and on the growth potential of selected market towns/service centres. These studies may be initiated by the DPU, and carried out by the DPU staff and other specialists. For these studies the database will be used with varying degrees of intensity, but supplementary fieldwork is often necessary.

The usefulness of the whole system is determined by the skills and experience of the DIDC/DPU staff in updating the information regularly and providing the departments at district, division and location level with the required data in the desired format. This again depends upon close co-operation between the DPU and the departments in the data collection, preferably with liaison officers in each department, and the thorough training of the DIDC staff in data collection, data analysis and data-retrieval methods. Formally, the DPU consists of a DDO, an assistant DDO, a statistical officer and a physical planning officer, together with supporting technical and administrative staff, including those attached to the DIDC. In practice, arrangements about the participation of the statistical officer and the physical planner still have to be made, and the training of DIDC staff is not only still in its initial stage but also biased towards librarian tasks with little attention being given to data collection and computer skills.

In sum, Kenya has embarked on a decentralization policy with a limited transfer of power and responsibilities to the District Commissioner and the heads of departments at the district level. The district planning process is co-ordinated by a single organization, namely the DPU, which functions as the technical support unit of the District Development Committee, with the DDO as the responsible officer. The transfer of authority also includes financial aspects. The planning process is strongly connected with district development plans in the form of blueprints and related annual expenditure plans. Monitoring and evaluation receive increasing attention, chiefly by means of inspection tours at irregular intervals by a large number of technical officers, namely, the heads of departments, under the supervision of the District Commissioner. The decentralization of planning contains a database element in the form of a DIDC, which is primarily seen as a reference library but at present the possibilities for the incorporation of a computerized database into a gradually increasing number of DIDCs in the country are being explored. The experiences in Kakamega District constitute an important input to these explorations.

Integrated rural development programmes in Sri Lanka

Government policy

Because of the stagnation in Sri Lanka's process of economic growth during the 1960s and early 1970s, the newly elected conservative government introduced an economic reform programme in 1977. Characteristic of its policy are a reduction in state intervention, a bigger role for market forces in the economic process and ample facilities to stimulate the activities of foreign enterprise (Richards and Gooneratne, 1980; Rajapatirana, 1988). The reform programme aims at improving the balance of payments, growth of the gross national product (chiefly through the promotion of private investment) and increased government expenditure on physical infrastructure to enhance the efficiency of private investments. Prominent large scale government projects include the Mahaweli energy-cum-irrigation project, the urban housing programme, the improvement of the state-owned plantations and the infrastructural facilities for the special economic zone. In this zone industrial growth is pursued through the provision of facilities to foreign firms, including tax holidays, attractive wage levels and easy financial conditions for the repatriation of profits.

The policy has led to accelerated economic growth, to an increase in exports and to self-sufficiency in food production. (For a detailed review of Sri Lanka's agricultural policy, see Thorbecke and Svejnar, 1987.) However, it has also caused increasing socioeconomic inequality, growing landlessness and deterioration of Sri Lanka's well-known system of social and community services. In addition, the expansion of foreign-owned industry has been limited and has not created the desired 'second Singapore'. Since 1983, the violent ethnic conflict, with accompanying high expenditure on defence, has not only increased Sri Lanka's foreign debts and dependency but has also ruined the economic and political stability needed to attract foreign investment.

Sri Lanka's new regional development policy started in 1978 with the appointment of one Member of Parliament from each district as a District Minister charged

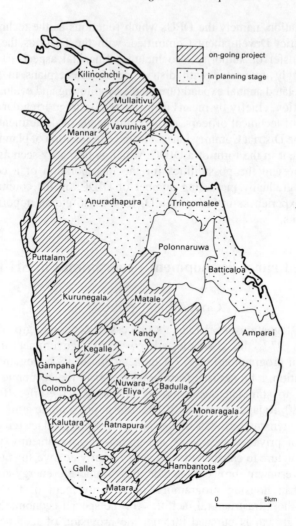

on-going project

in planning stage

Kilinochchi

Mullaitivu

Vavuniya

Mannar

Anuradhapura

Trincomalee

Polonnaruwa

Puttalam

Batticaloa

Kurunegala

Matale

Kandy

Amparai

Kegalle

Gampaha

Nuwara-
Eliya

Colombo

Badulla

Kalutara

Ratnapura

Monaragala

Galle

Hambantota

Matara

0   5km

Figure 13.3 Sri Lankan districts under the Integrated Rural Development Programme
(*Source:* Ramakrishnan, 1988)

with the task of designing a district development plan. In each district, a planning
unit was established for the necessary technical and administrative assistance,
under the co-ordination of the Ministry of Plan Implementation. A special so-called
decentralized budget is made available annually with a standard amount per
electorate for activities to be selected in close consultation with the Member of
Parliament of each electorate. In 1980 development councils were introduced at
various levels of administration to stimulate participation by the local population
in the planning process. The most important of these, the District Development
Council, receives separate financial means to carry out projects decided upon in
council meetings. At lower levels of administration, the Assistant Government
Agent or AGA division, and the *gramasevaka* development councils, have also

been established.

Finally, from 1979 onwards, Integrated Rural Development Programmes (IRDPs) are undertaken in a gradually expanding number of districts with technical and financial support from a wide range of donor agencies (Figure 13.3). The objectives of the IRDPs are phrased in general terms: they focus on the growth of production, employment and income with special attention to low-income groups, together with an improvement of the district planning process. These general objectives allow donor agencies to specify aims and approaches in accordance with their own policy preferences. Consequently, some IRDPs follow a blueprint approach with heavy investments in physical infrastructure, whereas others are characterized by a process approach with ample attention devoted to the improvement of employment, income and community services for the lower-income groups, and a strong emphasis on local participation. In addition, in many districts, separate bodies are established for management and administration of the IRDPs, leading to a dual system of decentralized regional development planning. Furthermore, co-ordination at the national level concentrates on contacts between individual IRDPs and sectoral ministries, and hardly so among IRDPs themselves (ILO/ARTEP, 1983; Ramakrishnan, 1988).

It is not surprising that such a broad framework for regional development pays little attention to standardized instructions for data collection and uniform data systems. Each IRDP operates its own procedures for problem identification, project formulation, priority setting and monitoring. Only after an initial period of experiments did the Ministry of Plan Implementation decide to establish a Monitoring and Evaluation Unit to streamline physical and financial monitoring and to issue provisional policy and planning guidelines to bring some standardization in planning procedures. The Scandinavian countries and The Netherlands render donor support for the co-ordination of the work on information systems for decentralized planning.

Nuwara Eliya Integrated Rural Development Programme

The Nuwara Eliya IRDP is a Netherlands-supported project that aims to promote social and economic development and to improve the planning process in the Nuwara Eliya District. The emphasis is being placed on the identification of, and financial support to, concrete activities to improve the production and living conditions of the low-income population. IRDP planning is characterized by a process approach, the implementation of activities through the line departments, the pursuit of a package approach and the spatial integration of activities in concentration areas. Project proposals are presented by line departments and other agencies, supported to varying degrees by IRDP planning officers. These proposals differ substantially in terms of underlying information, dependent on sector, type of activity and agency involved in the implementation. In any case, the implementation of measures to improve the present situation receives priority over the collection of basic data for both the IRDP and the implementing agencies. This priority is consistent with the objectives and with the process approach adopted in Nuwara Eliya.

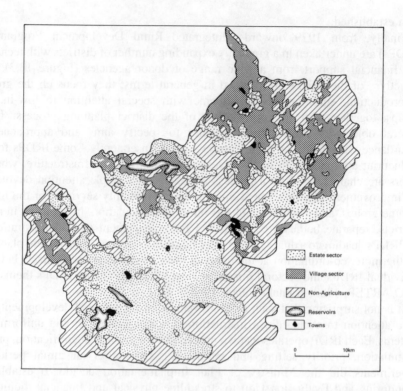

Estate sector

Village sector

Non-Agriculture

Reservoirs

Towns

0   10km

Figure 13.4   General land use map: Nuwara Eliya District (*Source:* Westeijn, 1989)

Nuwara Eliya District, with an area of 1,700 km² and located in the central
highlands of Sri Lanka, varies in altitude between 500 and 2,000 m. This variation
causes wide differences in temperatures and precipitation so that agro-ecological
conditions differ considerably from one part of the district to another. Almost 70
per cent of the cultivable area is under tea, chiefly on state-owned estates. These
estates make use of modern technology, have a relatively high output level and
a hierarchically structured labour force. The plantations have little economic
relationship with the so-called village sector, which is dominated by small-size
production units of different types, operating at rather low levels of technology
and chiefly working with household labour. Altitude, land type and farm size
influence the cropping pattern, and in this way the mono-cropped tea estates are
surrounded by a highly varied mosaic of farming systems, consisting of various
combinations of irrigated rice growing, homestead gardening, a cash crop (tea,
tobacco or vegetables) and dryland food cropping (Figure 13.4).

This dual economic structure is reflected in the district population which numbers
600,000, half of them Tamils living and working on the estates, the other half
Singhalese and living in the villages. Non-agricultural sources of employment and
income are scarce, which also explains the low degree of urbanization: only 6
per cent of the population lives in relatively small urban centres, which provide
mainly commercial and administrative services for the surrounding rural areas.

The district also has a rather low level of social and community services compared with the national average: a limited number of improved water-supply systems, many dilapidated and overcrowded school buildings and serious deficiencies in health facilities with regard to buildings, personnel, equipment and drugs.

The district database in Nuwara Eliya

The process approach adopted in Nuwara Eliya IRDP gives priority to the implementation of activities addressed to the easily observable and most urgent development needs. In addition, it is planned to combine the experiences gained in project implementation with a base-line survey for the identification of poverty groups, priority fields of intervention and geographical locations for concentrated improvement efforts. However, implementation has absorbed all attention and manpower, and only gradually has a rather superficial monitoring exercise emerged. This is carried out at irregular intervals, initially without much demonstrable impact on the selection of activities and areas of concentration. Over the years the monitoring system has been improved and the basis of record keeping changed from files via plan boards to a computerized system. In addition, the type of information collected has been standardized and regular meetings enhance the opportunities to use monitoring for the improvement of project performance. Finally, in 1987 (i.e. some seven years after the beginning of the project), a start was made with setting up a database, in addition to the existing monitoring system. Factors contributing to this are the expansion of qualified personnel, the availability of a micro-computer and the criticism ventured in an evaluation report about the absence of a base-line study, which hampers the assessment of the results of the project activities, and the general increase in interest for more detailed data at the co-ordinating ministry.

The database now established at the IRDP office in Nuwara Eliya contains or will contain

1. basic data for all units of the lowest level of administration (the *gramasevaka* or GS division) in order to select areas of concentration according to objective criteria;
2. detailed data for project identification in selected GS divisions;
3. sectoral data for priority sectors with an emphasis on physical infrastructure and community services (schools, health facilities and water supply);
4. monitoring data for ongoing project activities; and
5. a separate data set on selected estates and estate divisions.

The geographical units incorporated into the system are the GS division, the Assistant Government Agent (AGA) division and the district, which reflect the three levels of administration. The data are collected in close co-operation with the line departments responsible for the various sectors but the co-ordination is in the hands of the IRDP planners. In most cases joint teams are formed in each sector.

The data collection exercise is a long term process, commencing with easily available, reliable data, and only gradually moving towards more complex types of information with priority for the project areas. Data refer to the rural areas only; the urban areas will be left aside for the time being. Data collection for the village and estate sectors is undertaken separately. Although the estates area is incorporated formally into the territorial subdivision of the GS and AGA divisions, administration of the estates is executed almost entirely by the estate boards. Amenities, such as housing, water supply, sanitation, and health services, are provided by the boards without involvement of the district administration. Furthermore, certain data categories are not relevant to the estate sector (for example, land holding per household), while information is also more readily available in the estate sector as a result of a more sophisticated and well-kept system of administration. Finally, the estates operate a system of detailed and quite reliable data, which is extremely useful for planning purposes.

Data are stored in the IRDP micro-computer at the IRDP office – although not officially assigned with comprehensive district planning and although not formally responsible for the production of an overall district plan – is the best-equipped office in the district for such an exercise in terms of qualified personnel, transport, equipment and financial means. It maintains intensive relations with almost all line departments and the estate boards as a result of the co-operation in project activities.

Some common problems in the establishment of databases for district planning

The Kakamega and Nuwara Eliya cases point to substantial differences in development policy and, more especially, in decentralization policy between Kenya and Sri Lanka. The uniform process of top down decentralization in Kenya with strong co-ordinating roles for the District Commissioner, the DDC and the District Planning Unit contrasts sharply with Sri Lanka's lack of uniformity between districts as a result of donor preferences and priorities, and within districts as a result of the co-existence of several development budgets with their separate planning organizations. In addition, the two districts have sharply different economic structures. Although an internal differentiation in farm structure in association with agro-ecological variation, a low degree of urbanization and land shortage stimulating outmigration are phenomena common to both districts, the overriding dichotomy between village and plantation sectors in Nuwara Eliya heavily influences the district planning procedures and the nature of the database. Finally, the objectives of the two projects differ greatly: the Kakamega project is primarily geared to establishing a district information and documentation centre; the Nuwara Eliya IRDP gives priority to the planning and financial support of concrete development activities.

In spite of these differences, a number of common problems can be identified in relation to the experiences gained with establishing a database in the two districts. These problems are the lack of reliable key data, the use of different subdivisions/planning areas by the various line departments, the frequent changes in basic units of administration and planning over time, the absence of an effective monitoring system in the database and the regular administration's lack of experience in the establishment and use of computerized data systems.

The lack of reliable data for planning is a much-ventured complaint but nevertheless still valid, in spite of the frequent data collection exercises by government institutions, universities and individual researchers. In addition, it is still true that a substantial amount of information is not relevant for development planning, that potentially useful information becomes available late and that not all urgently needed information is accessible to planners. Major deficiencies observed in both districts are the absence of crucial information on production and production organization, such as yields in agriculture for agro-ecological zones; main characteristics for small non-agricultural establishments; information on employment and income; and the limited reliability of information with regard to socioeconomic differentiation because of their political sensitivity. This last category includes information on land ownership, share-cropping and food stamps. A good example of data deficiency is shown by the Agrarian Service Centres in Sri Lanka. Here detailed records are kept about agricultural production and yields but simple additions prove to be incorrect and political objectives influence data in such a way that the desired annual production increases are shown.

In addition to this general problem of data shortcomings that influences any type of development planning, regional planning is hampered more particularly by the lack of uniformity in the geographical units to which the data refer. Many line departments in both Kenya and Sri Lanka have their own territorial subdivisions that often differ from the administrative subdivisions used by the planning organizations. In the Kakamega District, data on education refer to school inspector zones, those on tea production are grouped together for collecting points and those on coffee input supply and production are aggregated for co-operatives and pulperies. In addition, regular data collection on farms by the Central Bureau of Statistics is organized through clusters that have an irregular distribution over the district, and do not relate to agro-ecological zones, whereas only aggregate results come available for the district level. In Nuwara Eliya planners involved in establishing a database are confronted with an even more complex pattern of territorial units. Apart from the estate village dichotomy with individual estates crossing GS division boundaries, line departments operate widely different subdivisions for the village sector. For example, the Department of Agrarian Services subdivided the district into 22 Agrarian Service Centre areas and, again, into 141 cultivation officer divisions. The Department of Agriculture operates with 3 segments, 22 agricultural instructor ranges and 59 extension officer areas. In complete contrast to these, the Ministry of Health organizes the health care system through 3 Medical Officer of Health offices, with 18 Public Health Inspector ranges and 110 Family Health Worker areas. Each line department collects and keeps data according to its specific territorial units, usually without taking account of the administrative subdivision in the district.

To complicate the work of the planners even further, these administrative subdivisions are not a stable system but subject to rather frequent changes. These changes are seldom inspired by the desire and opportunities for better planning, usually having a political motive and often in relation to an increase in employment opportunities for civil servants. In Kenya, new locations are created as soon as the present ones surpass a certain population threshold and subdivision usually takes place along the boundaries of sublocations. Here, the planners' problems may be solved by using the sublocation as the basic unit for data collection, whereas the location remains the lowest-level planning unit. In Sri Lanka a sudden increase in the number of GS divisions occurred in 1987 as a result of the change in function of the Special Service Officers (SSOs). Originally, each SSO served two GS divisions but as his duties were considered little different from the GS head, all SSOs were appointed heads and the number of GS divisions increased by one third. In Nuwara Eliya the number of GS divisions went up from 143 to 214 whereby the relation between the old and the new territorial units is not yet fully clear.

The incorporation of a monitoring system for ongoing projects and programmes into a district database creates a few special difficulties. The essence of the monitoring exercise may be described in general as the measuring, recording, collecting, processing and communicating of information to assist project management in decision-making. Experiences in the two districts, but especially in Sri Lanka, indicate that effective monitoring of development activities implies a clear formulation of the objectives of interventions and the accompanying time and payment schedules. In addition, it requires an unambiguous delimitation of responsibilities of the various parties involved in the activities and a detailed description of their tasks. In the context of regional development planning, projects and programmes are highly different, usually quite numerous and involving a large number of implementing agencies and beneficiaries. Yet the monitoring system must be operated by the limited personnel in the planning agency and with restricted financial means. It proves especially problematic to keep the system simple and to collect the necessary data at the appropriate time intervals. Even with the reliable data at hand, regular progress meetings were often evaded due to an understandable priority for implementation and the identification of new interventions. However, the effectiveness of the monitoring system is also reduced by the sociopolitical factors inhibiting the acknowledgement of delays and deficiencies in implementation, and the open discussion of remedial actions.

The use of micro-computers in developing countries is expanding rapidly, and also for planning purposes. However, the potential of micro-computers to facilitate data storage and data use for district planning is far from fully explored as yet. This potential use varies from the simple storage and retrieval of data to the application of methods for analysis and prognosis, specific planning techniques such as gravity models, shift and share analysis and the presentation of information in the form of tables, graphs, diagrams and maps. There are various factors responsible for this as yet suboptimal application, among which the non-availability of user-friendly software in developing countries or, if available, the absence of the necessary skills among planners, and the lack of knowledge as to how to use the software for planning purposes, combined with the shortage of training facilities, predominate.

Note

1. I am indebted to Gabriella Franck, Jan Hinderink, Henk Huisman and Wim Stoffers for critical remarks on an earlier version of this chapter.

References

Cohen, J. M. and Hook, R. M. (1986) *District Development Planning in Kenya*, Kenya Rural Planning Project, Ministry of Finance and Planning, Nairobi.

Conyers, D. (1983) Decentralization: the latest fashion in development administration, *Public Administration and Development*, Vol. 3, pp. 97–109.

Conyers, D. and Hills, P. (1984) *An Introduction to Development Planning in the Third World*, Wiley, Chichester.

House, W. J. and Killick, T. (1983) Social justice and development policy in Kenya's rural economy, in D. Ghai and S. Radwan (eds.), *Agrarian Policies and Rural Poverty in Africa*, ILO, Geneva, pp. 31–69.

Hyden, G. (1985) *No Shortcuts to Progress. African Development Management in Perspective*, Heinemann, London.

ILO (1972) *Employment, Incomes and Equality: A Strategy for Increasing Productive Employment in Kenya*, Geneva.

ILO/ARTEP (1983) *Planning for Rural Development. A Study of the District Integrated Rural Development Programme of Sri Lanka*, Bangkok.

Livingstone, I. (1986) *Rural Development, Employment and Incomes in Kenya*, Gower, Aldershot.

Rajapatirana, S. (1988) Foreign trade and economic development: Sri Lanka's experience, *World Development*, Vol. 16, no. 10, pp. 1143–57.

Ramakrishnan, P. S. (1988) Some aspects of the integrated rural development programme in Sri Lanka – past experiences and future perspectives, *Manchester Papers on Development*, Vol. 9, no. 1, pp. 136–58.

Republic of Kenya (1986) *Economic Management for Renewed Growth* (sessional paper no. 1), Nairobi.

Richards, P. and Gooneratne, W. (1980) *Basic Needs, Poverty and Government Policies in Sri Lanka*, ILO, Geneva.

Sterkenburg, J. J. (1987) Rural development and rural development policies: cases from Africa and Asia (Netherlands Geographical Studies no. 46, Department of Geography, University of Utrecht).

Sundaraman, K. V. (1987) An integrated approach to training for the establishment and use of information systems for subnational development planning, *Regional Development Dialogue*, Vol. 8, no. 1, pp. 54–70.

Thorbecke, E. and Svejnar, J. (1987) *Economic Policies and Agricultural Performance in Sri Lanka 1960–1984*, Development Centre of the Organisation for Economic Co-operation and Development, Paris.

Westeijn, (1989) Soil Conservation and Farm Improvement Programme IRDP, Nuwara Eliya, 5 August.

PART IV:
CONCLUSIONS

14

CONCLUSIONS AND PROSPECTS: WHAT FUTURE FOR REGIONAL PLANNING?

David Simon and Carole Rakodi

The changing context for regional planning

The 1980s have proved a difficult decade for almost the entire Third World and there is little sign of any significant improvement as we enter the 1990s. International and national forces have combined to create a very different environment for planners and planning from that prevailing during most of the 1960s and 1970s. At the same time, important advances have been made by social scientists in conceptualizing the relationship of planning to society and space. This has arisen out of ongoing theoretical debate as well as the apparent failure of many existing planning strategies (Chapter 1). The objectives of this book have been to survey these developments and to examine the extent to which the actual nature and practice of regional planning in a diverse selection of Third World countries have changed in response. In order to arrive at some coherent conclusions, we shall summarize each of the above sets of forces or issues in turn.

International forces

The current problems affecting the Third World were precipitated by the onset of global recession at the end of the 1970s. The increasingly severe debt crisis arose through the combination of several factors, including depressed world commodity markets and deteriorating terms of trade for Third World producers, vastly increased oil import bills for the majority of states that do not produce 'black gold', corruption and an increasingly conservative international political climate. Particularly important has also been the need to repay, in the face of real interest rates that have reached unprecedentedly high levels, injudicious foreign borrowing undertaken at flexible but low interest rates during the 1970s. Much of this capital was borrowed in order to continue with often grandiose and inappropriate development schemes, many of which formed elements of regional development programmes. Even most oil producers, represented by Ecuador and Indonesia in this volume, have not escaped this fate. Indeed, following more recent decreases in oil prices and for the other reasons just cited, their decline into debt during the 1980s was precipitate. Many of the world's most peripheral states, such as most Pacific islands, have long been heavily reliant on external aid flows, a situation unlikely to change in the foreseeable future. Global capitalism has certainly undergone a severe crisis but it is important to note that the crisis has embraced all modes of production, be they capitalist, state capitalist or socialist.

No less important, however, has been the diagnosis of the causes of the crisis by the North's major financial institutions and aid agencies, and their response. Structural adjustment lending and its associated conditions have forced governments to devalue their currencies, to cut state expenditure, including the reduction or removal of subsidies, to reduce direct state involvement in economic production and to promote exports of primary products in order to service their debts and pay for imports. The consequences have been reduced welfare provision for the poor, increased social and spatial inequality, greater exploitative pressures on the environment and, frequently also, deteriorating infrastructure.

National forces

Following independence, Third World governments almost invariably assumed an active economic role commensurate with their political commitment to overcoming inherited colonial underdevelopment and sociospatial inequalities. Irrespective of ideological orientation, governments were often the only agents able to command or co-ordinate the resources required for construction of large scale infrastructure and industrial plants. Furthermore, such investment was generally seen to have great strategic importance in terms of promoting national self-reliance and symbolic value indicative of the path to modernization. It also often provided a valuable tool for the centralization and consolidation of power by the new political élites. Optimism about the prospects for development based on industrialization was reinforced by global economic expansion during the 1960s and early 1970s, when foreign markets for Third World exports were buoyant in terms both of volumes and prices.

There was general confidence in the ability of the state to direct resource allocation by means of development planning in order to restructure the economy and society and promote greater equity. Regional planning acquired increasing recognition as part of this tool-kit, as witnessed by the incorporation of an explicitly spatial element into the five year plans of almost all countries during the 1970s. Yet, as discussed in Chapter 1, such efforts, whether geared to explicitly capitalist modernization or supposedly more socialist transformation, were generally state centred and top down, focusing on some variant of a growth pole or growth centre strategy, and based on an inadequate conception of the relationship between space and the political economy. In essence, regional planning has been a tool of the state in redistributing, or giving the appearance of redistributing, resources from the core towards the periphery through conspicuous investment. However, the top down nature of state efforts has stood in marked contrast to the increasingly bottom up character of much development activity undertaken by NGOs and local or regional social movements. This contrast is symptomatic of the fundamental dialectical contradiction between the substantive nature and legitimizing ideology of many states.

Given its greatly enhanced role, the state apparatus expanded beyond all recognition from its colonial predecessor. In many countries it acquired an unwieldy bureaucratic structure unsuited to the management of a modern economy, being beset by overlapping responsibilities and ill-defined tasks. Somewhat ironically, given the activist role of the postcolonial state in restructuring society, civil service structures inherited at independence have proved one of the most difficult elements to change. Aside from inertia, probably the most important reason for such resilience is that these structures, geared as they were to control of the indigenous majority by the colonial minority, have proved admirably functional to continued postcolonial centralization and entrenchment of a 'bureaucratic bourgeoisie'.

These trends have been thrown into reverse by the very different international conditions and pressures of the 1980s outlined above. Despite the rhetoric of 'accelerated development' and more recent variants emanating from the US government, IMF and World Bank in Washington, DC, the emphasis for virtually the entire decade has fallen on preventing collapse of the international financial system. Demands from Third World countries and progressive think tanks for a New International Economic Order are paid lip service at best by the powers that be, who argue that world development will best be promoted by the world's poorer countries being cajoled back into solvency.

Consequently, debt service has all too often taken precedence over positive development initiatives and the needs of the poor. Almost invariably, as a condition of structural adjustment lending, countries are being forced to curtail the role of the state and the size of its bureaucracy. Economic crisis has, in any case, left governments with little flexibility in resource allocation and forced them into crisis management rather than a developmental approach to policy and action. The administrative capacity of the state has not expanded commensurately with its size. An unwieldy bureaucracy, with ill-defined tasks and unclear responsibilities, has contributed to the failure of the state to deliver progress. It has, therefore, been easy for the new generation of economic theorists to blame the crisis on a misplaced reliance on the state rather than on the market to achieve development. Mal- or underdevelopment, it is argued, must be redressed by the state adopting essentially a facilitatory role with respect to private enterprise rather than being directly active in the economy. Consequently, the civil service has been cut, state corporations and some other activities have been wound up or privatized as part of the new emphasis on market forces, while investment has been redirected to facilitate production for export. Moreover, budgetary stringency has forced the curtailment of subsidies and other transfer payments that previously enabled an element of income redistribution, a cut in the real value of public (and other formal) sector wages and cutbacks in other so-called non-productive expenditures, such as the (re-)location incentive packages associated with regional planning strategies. Certain 'weak' forms of decentralization have been promoted, but the state in most Third World countries remains concerned with the retention of centralized political power so that little real empowerment of regional populations has occurred. Against this background, we now turn to a more detailed consideration of how spatial policy has changed in practice during the last decade or so.

Regional policy and the nature of the state

As argued by Simon in Chapter 1 and exemplified in a range of the country studies, one of the most important advances in conceptual and research terms has been increasing recognition that analysis of regional and other spatial policies requires an understanding of the nature of the state and social context. Examination of such strategies in isolation, now often dubbed 'spatial fetishism', is meaningless since space does not exist in isolation from other aspects of economy, society and polity. Space is also not an abstract or neutral dimension; we need a 'relational' conception of space. The way in which economic, social and political relations find spatial expression, and attempts by the state to manipulate space, are integrally related and dynamic. These are rightly the stuff of spatial science, which by definition is therefore a social science rather than the search for abstract laws and formulae.

A central feature of the activist state in the postcolonial era has been the high degree of centralization of effective power and decision-making. Under the right conditions, this is not without potential advantages, but centralization has generally been excessive. Whatever the rhetoric about people's power, local autonomy and the need to transform unjust colonial structures, few genuine attempts have been made to institute democratic decentralization. Common measures have been the establishment of regional offices of sectoral ministries, perhaps with some attempt at interministerial co-ordination. Even where subnational assemblies have been created, their role has generally been to advise or elect representatives to the central state. Their effective power or discretion over resources has usually been small. This is true of most of the countries covered in this book, including Costa Rica, Ecuador, Indonesia, Kenya and most Pacific islands.

Across the ideological spectrum, most conspicuously in Africa, the dominant tendency of centralization has often been symbolized by one-partyism. This has generally been justified on grounds of promoting national unity, avoiding the distractions and costs of multiparty politics and providing the strength and continuity essential to the daunting tasks of national development. This identification of party with state has, however, facilitated the suppression of opposition both within and outside the party, permitting the entrenchment of a ruling élite that, almost inevitably, has become less efficient, more greedy and totalitarian over time. Many other states have been subjected to periods of civilian or military dictatorship that, by definition, have been repressive and highly centralized regimes.

More generally, usurpation and sometimes brutal destruction of the wealth of nations – their people, environment and resources – have often been justified in the name of national security and development. This is not some historical anomaly; the list of contemporary Third World examples spans all ideological orientations and continents, from the Brazilian Amazon to South Africa's bantustans, the Kurdish regions of Iran, Iraq and Turkey, transmigration in Indonesia and Chinese-occupied Tibet. Against this background, the lack of domestic employment prospects and the increasing transnationalization of the world economy, international labour migration has become common. Although not explored in the present book, this phenomenon can have serious implications for regional inequality and development prospects, particularly where it involves a high proportion of specific skill, gender or social groups from peripheral regions.

The need for an understanding of the nature of the state furthermore suggests the need for caution in making broad generalizations. As James Sidaway and David Simon argue in Chapter 2, assertions about the nature of urbanization and spatial inequality under socialism have frequently failed to take adequate account of the heterogeneity of states claiming some form of socialism as official ideology, of differences between these forms and between ideology and practice or of the diversity of spatial forms found in these countries. The conspicuous paucity of critical research on territorial questions in such states, given their large number, only underlines the reliance of most social theorists on capitalist – almost exclusively advanced capitalist – societies in formulating new ideas on urban and regional development in the light of global capitalist restructuring. These are then all too commonly implied or held to be of universal applicability. Our understanding would be greatly enhanced by taking account of different sociospatial forms and associated constraints of capital scarcity and immobility. For it is the social, economic, political and spatial consequences of the respective policies, rather than the underpinning ideologies as such, that distinguish different modes of production.

These issues are clearly exemplified in Peter Slowe's longitudinal study of Guinea, which spans two contrasting periods. The first, often regarded as a variant of socialism, was state-centred development aimed at national integration compatible with at least some indigenous values, while the second has been overt capitalism seeking to maximize economic growth within the world economy. Somewhat ironically, the contradictory effects of the latter have thus far apparently been cushioned by the sense of national identity engendered during the earlier phase.

By contrast, most Pacific island states are only now establishing coherent national and subnational planning frameworks in an effort to give effect to constitutional commitments and the objectives of development plans. They face particularly severe constraints in terms of conventional development strategies, and the applicability of many imported theories and methods is questionable. Existing attempts at decentralization have been inadequate and ill conceived. However, Hidehiko Sazanami and Roswitha Newels argue that the underlying principles are the same, and they advocate a multilevel approach to development planning and management with an explicitly spatial element. This approach embraces concerted strengthening of institutional capacity at all scales, coordination and integration between national sectoral agencies and subnational bodies, together with maximum local participation in the process, utilizing indigenous channels where possible. Some of the experiences with spatial planning of other countries reported in this volume may well have relevance to the orientation of these new initiatives for island microstates.

Impact of the debt crisis and structural adjustment

Selective investment strategies

Although still evident in various guises across a range of countries, growth pole strategies have increasingly been criticized on a number of grounds (see Chapter 1). Three of the studies in this volume cast useful new light on the issues involved in concentrated investment strategies by examining recent changes. None of the strategies evaluated – production for export in remote regions (Costa Rica), agrarian and industrial modernization with particular reference to intermediate cities in Ecuador and export-processing zones in the very different states of Mexico and China – has succeeded substantially in reducing regional inequality, converting economic growth into positive development and thereby providing substantial benefits to the local poor. While providing some palliative against the worst effects of unemployment, the primary beneficiaries have been the state and owners of local, national and international capital.

In Costa Rica, the objective of reducing regional inequalities has been suppressed as a result of the debt crisis. However, Arie Romein and Jur Schuurman argue that there has not been a consequent resurgence of functional regional planning over territorial regional planning, because strategy has always been essentially functional in nature. Stella Lowder shows that, in Ecuador, state policy was also geared to the promotion of capitalist accumulation and thus did not promote the structural transformations required to ensure greater sociospatial equity and greater resilience to the external shock of the debt crisis. And as Leslie Sklair clearly demonstrates, the increasingly popular strategies that pursue export-led industrialization fuelled by foreign investment and technology, of which export processing zones are the conspicuous spatial manifestation, create relatively hi-tech enclaves integrated into the international capitalist economy rather than their host economies. They also reflect very graphically the pressures for countries to abandon earlier isolationist or self-reliant policies in the search for capital and technology. However, some evidence now emerging suggests that differences may exist in both the nature and consequences of transnational investment by firms originating in newly industrializing countries (and, indeed, in Japan) from that by North American and European-based corporations.

Such selective investment strategies have been associated with state directed development, which included a top down, technocratic approach to planning that has pervaded national economic, regional, urban land use and project planning alike. The unsuitability of end-state plans as a basis for action in conditions of extreme uncertainty, the inability of many states to implement grandiose investment schemes, together with the failure of such projects to achieve national or regional development objectives, led to attempts to change both the nature of political and administrative territorial organization and approaches to planning.

Decentralization and rural development

Decentralized rural and regional planning initiatives have been undertaken in a range of countries for reasons including concern at the rate and scale of urbanization, a desire to reduce regional inequality, the need to secure an adequate food supply and/or level of agricultural exports, the need to provide at least a semblance of resource redistribution and local responsiveness and a quest for more effective policy implementation. The precise strategies have changed somewhat over time as a reflection of changing state priorities and conventional wisdoms in planning. The earlier emphasis on blueprint plans, supposedly replicable nationwide, has sometimes shifted in favour of a limited number of integrated rural development schemes and/or the creation of additional regional organs and agencies. These are intended to increase the local capacity for integrated and participatory development planning and management. However, the responsibilities and roles of various actors generally remain inadequately co-ordinated and their powers too restricted relative to those of sectoral ministries. Carole Rakodi's conclusion (p. 147) that 'in practice, the strengthening of central state control has taken precedence over a political ideology that stresses the populist nature of the state' is applicable to virtually all the case studies from Africa and Asia covered in this volume. Marcel Rutten's study of district level development in Kenya exemplifies this particularly well. Substantive decentralization has never really been on the agenda, and the District Focus thus lacks clarity of purpose, there is little effective local participation or co-ordination between regional representatives of sectoral ministries and the budget preparation process is still centrally directed and approved. The sort of bureaucratic consultation exercises in which planners seek inhabitants' views on their proposals or plans fall far short of true participation.

In Kenya and elsewhere, several recent changes attributable to the debt crisis and increased reliance on external resource transfers warrant mention. First is the reduction of state autonomy in establishing and executing policies. Most bilateral and multilateral aid donors now set firm criteria for granting resources, monitoring their use and the effectiveness of the outcomes. Very often, technical and other personnel from the donor country or organization are involved on the ground throughout. Depending on the precise arrangements, this can indeed represent a positive development. However, in several cases studied here (e.g. Zambia and Sri Lanka), the extent of aid reliance and the spatial concentration of schemes funded by different donors has had serious consequences for domestic policy. Since each donor has its own particular sectoral preferences and set of procedures, and has concentrated its activities in one or two regions to maximize its own efficiency, overall national co-ordination has become even less coherent than previously. Donor and project proliferation has led to administrative overload in recipient countries and financial overload on domestic recurrent budgets. Planning has consequently often been reduced to the matching of donor preferences to projects needing funding, even if these projects are not high in national development priorities. More than one observer has suggested that Zambia has as many agricultural policies as foreign donors. Under such circumstances, regional inequality is likely to increase, whatever the impact of aid projects in terms of output, thus counteracting the state's spatial policies. If sustained, this situation could ultimately undermine the legitimacy of the state, and the mere fact that it has been tolerated indicates the predicament facing countries that are increasingly aid dependent. How such developments are affecting the peasantry is difficult to ascertain and depends very much on the region and country-specific conditions. Dependence on aid is further complicated in many countries by the growth in NGO involvement, although none of the chapters in this volume addresses this phenomenon directly. NGOs may be successful initiators of small scale projects that directly benefit poor local people, but these organizations are diverse and frequently not amenable to integration into national, regional or local administrative processes.

The conditionality attached to IMF/World Bank loans has had an even more profound impact across all sectors. Resources available for explicitly spatial policies or discretionary regional development have been savaged or cut altogether, while the urban poor have borne the brunt of reduced subsidies on staple foods and other necessities. Conversely, rural peasant producers may actually have benefited from higher producer prices if market structures operate remotely efficiently. There is also now some evidence that the rolling back of the centralized state might enhance the possibility for greater regional organization, decision-making and autonomy in some countries.

Bill Gould's study of Ghana is of particular interest because of that state's extreme plight, having long been regarded as a basket case, and because it is arguably the country furthest down the road of change induced by structural adjustment. Yet even here, evaluation is premature because of the recency of the decentralization initiatives stimulated by the World Bank as part of its second-stage adjustment lending. In contrast to other countries, the state has now embraced the programme in a positive manner, and Gould holds out the prospect of both quantitative and qualitative improvement in the educational and other sectors. However, this will depend on the state's ability to reconcile the political pressures for central control with its administrative commitment to decentralization. It will also require a continuation of the massive external resource inflow coupled with appropriate macro-economic policies to ensure that the later switch to internal cost recovery does not simply reproduce earlier patterns of inequality. Explicitly spatial policies alone will achieve little.

Allert van den Ham and Ton van Naerssen make the same point with respect to Indonesia, a country with a highly centralized state apparatus and some of the most extreme imbalances in social and spatial development. A greater role for the regions seems inevitable both because of the stark contrasts between Inner and Outer Islands, and because of the effective suspension of the government's directed regional policy due to lack of funds as oil revenues dry up. The centrepiece of this policy has hitherto been the problematic transmigration programme. Changes over the last decade have increased the possibilities for bottom up planning, but concrete progress has been only tentative so far. The crucial factor will not ultimately be the mere existence of decentralized powers and planning procedures but the extent to which regional authorities are prepared to tackle problems of poverty and resource distribution. This will depend in turn on the political and economic class structure of these authorities.

Botswana is one of only a handful of Third World countries enjoying continued economic buoyancy and initiating increasingly positive development policies. Furthermore, given the country's small population, vast size and democratic political system, area-based decentralization is both important and highly appropriate for study. Basic district-level planning has been found inadequate, hence the introduction of complementary initiatives at subdistrict level via the Communal First Development Area strategy and an improved land use planning experiment in Southern District. Although lack of political and bureaucratic commitment at the centre is hindering progress, Ruud Jansen and Paul van Hoof see great potential in these initiatives because of their process orientation with provision for intensive grassroots consultation and their directed yet integrative approach.

Associated with some recent attempts to develop local indigenous capacity to formulate and implement policies, programmes and projects we can thus perceive the beginning of a shift to adaptive, process-oriented participatory planning. This is being accompanied by a change of emphasis from earlier preoccupations with physical aspects of regional development, such as settlement patterns and infrastructural investment, to greater concern for understanding the organization of production and ensuring compatibility between national and regional economic objectives. Lessons from implementational failures are likewise being learnt and embodied in new approaches to the decision-making and administrative process, such as those Botswana, Kenya and some donors in Zambia and Zimbabwe are trying to develop.

Jan Sterkenburg's experience in Kenya and Sri Lanka provides interesting comparative insights into the practical problems involved in setting up databases and monitoring procedures appropriate to decentralized spatial planning. Foremost among these are the availability, reliability and appropriateness of both quantitative and qualitative data, lack of interministerial co-ordination even on matters as basic as the delimitation of spatial planning units, frequent changes to such basic administrative units and the absence of ongoing monitoring systems. The last mentioned, together with data analysis, presentation and projection, can be facilitated by use of micro-computers, use of which is increasing. Foreign aid programmes have been instrumental in introducing this technology to many planning institutions and agencies in Third World countries as part of wider decentralized development assistance programmes. While potentially appropriate tools for progressive and flexible local planning initiatives, hurdles to their more widespread and effective adoption are presented by the general lack of suitable software, the paucity and rapid turnover of trained personnel and, in some places, lack of reliable electricity supplies. A word of caution is also necessary, however, since there is always a risk that new technology will be used to entrench technocratic planning. This is the very opposite of what is being advocated.

The road ahead

Several important points emerge from the foregoing. They are not only conclusions in the conventional sense but key considerations that will shape both regional development policy in practice and our conceptualization of it.

First, structural adjustment and the new aid conditionality are here to stay for the foreseeable future and cannot simply be wished away. However distasteful this may be, Third World governments are being compelled by force of circumstance to bite the bullet. Even a country like Zambia, which has broken off negotiations with the IMF on more than one occasion because of the high domestic political cost, has been unable to do more than postpone the inevitable.

Second, comprehensive restructuring programmes in the vast majority of Third World countries will need to be supported by large inflows of foreign resources. Greater donor involvement is thus on the cards, with possible consequences for decision-making sovereignty by recipient states. Under the right conditions, foreign supervision and technical assistance can be beneficial to those most in need, but very often they serve the interests of dominant groups and/or the state.

Third, most decentralization or spatial planning schemes hitherto promoted by national governments and the international technocratic/donor corps have been 'weak' in the sense of attempting to treat only the symptoms rather than the underlying causes of inequality and poverty. This reflects a central contradiction between the highly centralized nature of most Third World states and the supposedly populist ideologies from which they derived and have sought to perpetuate their legitimacy. Simply creating subnational spatial entities, assemblies or planning bodies is an empty gesture in the absence of political will and the conferral of powers and resources commensurate with their duties.

This brings us to the fourth point, namely, that space cannot be treated in isolation from other facets of economy, society and polity. Most explicitly spatial policies have proven ineffectual while it is well known that functional policies, such as macro-economic measures, often have major and unequal spatial impacts. We know, too, that the continued widespread deployment of regional policy has been functional to the state and allied segments of capital. Greater understanding of the need to integrate a spatial focus with associated measures geared to tackling the underlying problems of poverty, unequal resource distribution and environmental degradation, holds the key to future advances in regional development policy.

Fifth, tackling poverty and inequality demands more than handouts. It requires empowerment of the poor, access to the bases for accumulating social, economic and political power. This is a radical task, the essence of 'strong' basic needs strategies, whether defined in space or not. And by definition, this is inimical to the interests of a centralized state and allied economic forces. Space is but one dimension in which poverty and inequality are manifest. They occur at all scales and in all sectors, not least in terms of gender and intergenerational relations within households, and in both rural and urban areas. Policies need to be focused on unempowered groups defined in terms of relevant criteria, including ethnicity, class, caste, religion, gender, age and locality. An appropriately framed spatial focus can form a valuable element in integrated 'strong' development strategies, which must above all be poverty focused, socioculturally compatible and environmentally sustainable. The existence of sufficient political commitment by the state to operationalize or at least not to undermine any such strategy is, of course, a prerequisite. Many of the studies in this volume highlight necessary components of 'strong' strategies, both territorial and functional. Important among them are

- clearly focused aims, objectives and target areas;
- substantial grassroots participation and control over decision-making throughout, as distinct from mere consultation by planners;
- availability of appropriate resources, training and powers;
- orientation towards planning as a process rather than towards the plan as a product;
- ongoing monitoring and evaluation of qualitative as well as quantitative dimensions, and in terms of appropriate criteria;
- integration between sectors and ministries to enhance the effectiveness of state activities;
- increased prosperity without sacrificing sustainability; and
- replicability without loss of local appropriateness and accountability.

The implementation of such 'strong' strategies requires consideration of the bureaucratic environment within which regional development planning must take place, as well as an understanding of the changing national and international context. This involves an understanding of the nature of the state apparatus, including the administrative system and the economic and political interests expressed through other mechanisms, both formal political structures and informal systems of influence and patronage. The imperatives to which the administrative agencies involved in regional development respond will inevitably shape the environment in which policy is determined and implemented. An approach to regional development that addresses inequalities, meets basic needs and achieves greater prosperity and sustainability, requires not only appropriate decision-making structures and sensible policies but also compatible administrative environments and practices.

Finally, there is some cause for optimism amid the gloom and crisis. Perhaps paradoxically, given the obstacles discussed above, present circumstances may lend themselves better than hitherto imagined to the more widespread adoption of just such strategies. State socialism in the Third World has been rolled back almost everywhere, and the dramatic changes now sweeping Eastern Europe will undoubtedly have a major impact on the remaining adherents, both genuine and notional, of 'Marxist-Leninism'. Although it is too early to judge with confidence, these changes appear to signify a return to capitalism in several countries rather more than a progression from state to democratic socialism.

Nevertheless, most centralized capitalist states of the 'South' have fared no better and also face dramatic restructuring. Irrespective of official ideology, many Third World states face not only a crisis of solvency but also one of legitimacy. If the political will exists or can be created, the search for alternatives now beginning may therefore well represent an historic opportunity for the much vaunted slogan of 'power to the people' to acquire substance beyond mere rhetoric. Many NGOs and social movements, often operating in partnerships that link Third and First Worlds, are already providing workable and replicable lessons in promoting small scale, local-level, bottom up development. In addition to its redistributive role at the national level, in reallocating resources from the better endowed or more developed regions to resource poorer or more underdeveloped areas, regional planning, as defined here, could also become an important tool in facilitating such local initiatives, thereby enabling self-reliant and sustainable development. This may ultimately have significant potential to advance the prospects for a New International Economic Order from the bottom up.

INDEX